T0212789

# Measuring SIP Proxy Server Performance

Sureshkumar V. Subramanian • Rudra Dutta

# Measuring SIP Proxy Server Performance

 Springer

Sureshkumar V. Subramanian
Cisco Systems, Inc
Raleigh, NC, USA

Rudra Dutta
Department of Computer Science
North Carolina State University
Raleigh, NC, USA

ISBN 978-3-319-34575-8          ISBN 978-3-319-00990-2 (eBook)
DOI 10.1007/978-3-319-00990-2
Springer New York Heidelberg Dordrecht London

Printed on acid-free paper

Springer is part of Springer Science+Business Media (www.springer.com)

*To my parents for their guidance and affection*

*To all my teachers, lecturers, and professors for sharing their knowledge and intelligence*

*To my wife, Gayathri, for her endless love, encouragement, and motivation*

*To my kids, Shobana and Shravan, for their understanding and support*

–Suresh

# About the Authors

Sureshkumar V. Subramanian is currently working as a Senior Software Engineer in the Systems Development Unit (SDU) at Cisco Systems in Research Triangle Park, North Carolina. He has worked in various Software System development engineering roles in telecommunication and networking technology companies such as NEC America Inc., Tekelec Inc., Nortel Networks, Lucent Technologies, and Cisco Systems for the past 23 years. Suresh presented and published technical research papers in IEEE conferences related to SIP proxy server performance. He received his Ph.D. in Computer Science from North Carolina State University. He lives with his wife and two children in Raleigh, North Carolina.

Rudra Dutta was born in Kolkata, India, in 1968. After completing elementary schooling in Kolkata, he received a B.E. in Electricial Engineering from Jadavpur University, Calcutta, India, in 1991; a M.E. in Systems Science and Automation from the Indian Institute of Science, Bangalore, India, in 1993; and a Ph.D. in Computer Science from North Carolina State University, Raleigh, USA, in 2001. From 1993 to 1997 he worked for IBM as a software developer and programmer in various networking-related projects. From 2001 to 2007, he was employed as Assistant Professor, and since 2007, he has been employed as Associate Professor, in the department of Computer Science at the North Carolina State University, Raleigh. During the summer of 2005, he was a visiting researcher at the IBM WebSphere Technology Institute in RTP, NC, USA. He is married with two children and lives in Cary, North Carolina, with his family. His father and his sister's family live in Kolkata, India.

# Preface

The wide-scale deployment of Internet combined with several advancements in hardware and software technologies created opportunities for several Internet-based applications such as Voice over IP (VoIP) that involves the delivery of voice, video, and data to the end user. In recent years, Internet Protocol (IP) telephony has been a real alternative to the traditional Public Switched Telephone Networks (PSTNs). The IP telephony offers more flexibility in the implementation of new features and services. The Session Initiation Protocol (SIP) is becoming a popular signaling protocol for Voice over IP (VoIP)-based applications. SIP is a peer-to-peer multimedia-signaling protocol standardized by the Internet Engineering Task Force (IETF) that plays a vital role in providing services to IP telephony. The SIP proxy server is a software application that provides call routing services by parsing and forwarding all the incoming SIP packets in an IP telephony network. The efficiency of this process can create large-scale, highly reliable packet voice networks for service providers and enterprises. We established that the efficient design and implementation of the SIP proxy server architecture could enhance the performance characteristics of a SIP proxy server significantly. Since SIP proxy server performance can be characterized by its transaction states of each SIP session, we emulated the $M/M/1$ performance model of the SIP proxy server and studied some of the key performance benchmarks such as average response time to process the SIP calls and mean number of SIP calls in the system. We showed its limitations and provided an alternative analytical solution based on $M/D/1$ model in our first part of the research. Later, based on the latest advancement in software engineering and networking technologies, we studied another $M/M/c$-based SIP proxy server performance model with enhanced performance and predictable results. Then, we studied the performance and scalability of the SIP proxy server, when the calls are routed through local and wide area networks. Finally, we expanded the research further to evaluate the performance impacts of the SIP proxy server when secured and non-secured transport protocols are used to transport the packets, since the deployment and delivery of SIP value-added services in the public network carries significant security issues.

Raleigh, NC, USA                                    Sureshkumar V. Subramanian
Raleigh, NC, USA                                                    Rudra Dutta

# Acknowledgements

The success of an individual is a collaborative effort of many people and I am not an exception. First, I am very grateful to Dr. Rudra Dutta for patiently listening to my ideas, raising creative questions, bringing up good discussion points with rightful criticisms, making me think more, pointing me in the right direction, and solidifying them into a coherent material that you are about to witness. Heartfelt thanks to Cisco Systems for their support for writing this book. None of this is possible without my wife, Gayathri. She encouraged me throughout these years with her endless love and dedication. Thanks to my kids for adjusting to my tough schedule with my office work and other writing activities. Finally, I am grateful to my parents for raising me with high values and constantly reminding me the importance of education in life.

Sureshkumar V. Subramanian

# Contents

# List of Figures

# List of Tables

# Acronyms

| | |
|---|---|
| ACM | Address Complete Message |
| AIN | Advanced Intelligent Network |
| ANM | Answer Message |
| AOR | address or record |
| BHCV | Busy Hour Call Volume |
| B2BUA | Back-to-back User Agent |
| CHT | Call Hold Time |
| CCITT | International Telegraph and Telephone Consultative Committee |
| CCSN | Common Channel Signaling Networks |
| CGA | Common Gateway Interface |
| CPL | Call Processing Language |
| DoS | Denial of Service |
| DNS | Domain Name Service |
| FQDN | Fully Qualified Domain Name |
| HTTP | Hypertext Transfer Protocol |
| IMS | IP Multimedia Subsystem |
| IN | Intelligent Network |
| IP | Internet Protocol |
| IAM | Initial Address Message |
| IETF | Internet Engineering Task Force |
| LAN | Local Area Network |
| LDAP | Lightweight Directory Access Protocol |
| MGC | Media Gateway Controller |
| PDA | Personal Data Assistant |
| PSTN | Public Switched Telephone Network |
| QoS | Quality Of Service |
| RLC-AM | Radio Link Control acknowledgement mode |
| RLP | Radio Link Protocol |
| REL | Release Message |
| RLC | Release Complete Message |
| RTP | Real-Time Protocol |

| SCP | Signaling Control Point |
| SDP | Session Description Protocol |
| SIP | Session Initiation Protocol |
| SRTP | Secure Real-Time Protocol |
| SS7 | Signaling System No.7 |
| TCP | Transmission Control Protocol |
| TLS | Transport Layer Security |
| UAC | User Agent Client |
| UAS | User Agent Server |
| UDP | User Datagram Protocol |
| URI | Uniform Resource Identifier |
| VoIP | Voice Over IP |
| WAN | Wide Area Network |

# Chapter 1
# Introduction

Internet has evolved from network architectures in which, general connectivity was far more important than consistent performance, and the Internet of today consequently exhibits less predictable performance. Generally, the delivery of Internet Protocol (IP) services is designed to be "Best effort" and the packets were treated equally. In case of Public Switched Telephone Network (PSTN), routing is very expensive due to higher bandwidth available for high performing and reliable interactive audio and video sessions. Whereas, in IP Networks, the routing is reasonably affordable but sharing the resources makes the IP services a real challenge. There are some fundamental issues in the audio and video services like the latency, packet loss and jitter. The basic principles of PSTN and Internet are different. The PSTN is a network that transports only the voice media with the most reliable delivery mechanism. Internet on the other hand can transport any media (voice, video and data) using the best effort delivery mechanisms. PSTN requires a simple endpoint (simple phone) and an expensive switch to transport the voice circuits along with other services. Internet uses the routers to receive the packet; forwards the packet to the next router and so on until the packet reaches the intended destination. The power of Internet is concentrated on the edges of network elements such as powerful desktop, laptop computers, cell phones and other digital hand held devices (Gurabani 2004).

Services are viewed as the greatest challenge among Internet based telecommunication networks (Gurabani 2004) and the end users expect the Internet telephony service in par with PSTN services. Internet telephony services are more controlled by its endpoints such as Personal Computers (PCs), Personal Digital Assistants (PDAs), and cell phones, whereas services in PSTN are centralized in a central office switch. To compare all the PSTN services such as 1–800 services, 911 emergency services etc. with the Internet telephony endpoints in VoIP networks, we need to evaluate the performance, scalability and reliability of VoIP network (De et al. 2002). In recent years, SIP, an Internet Engineering Task Force (IETF) standard has been considered as a promising signaling protocol for the current and future IP telephony services because of its simplicity, flexibility, and built in security features (Lennox 2004). If IP telephony along with SIP signaling is to be the modern

S.V. Subramanian and R. Dutta, *Measuring SIP Proxy Server Performance*, DOI 10.1007/978-3-319-00990-2_1,

day replacement for PSTN, it should meet same level of Quality of Service (QoS) and security. There are several ongoing discussions on the QoS of IP telephony services and SIP within the IETF and other research communities. Several Internet telephony events such as initialization of a call (making a VoIP call), arrival of call, parsing the headers, location updates such as location registration and de-registration in case of cellular networks, routing the call are the performance characteristics that needs to be studied. These key functions are designed and implemented in the SIP proxy server software, which forms a central part of the control plane corresponding to the VoIP data plane.

Traditionally, a Poisson process is used to characterize PSTN call arrivals. For sizing of the system, generally Busy Hour Call Volume (BHCV) is used as a requirement, which indicates the arrival rate during the busiest hour of the busiest day. The time between the setup and tear down is the conversation time or Call Hold Time (CHT). The exponential distribution is the traditional method of approximating the CHT in a switch. Since, the SIP proxy servers maintain transaction state, not the call state, call duration time is generally ignored. More important for SIP proxy servers is the time between call arrival and answer, as it determines the number of transactions that the server needs to maintain. Very few research works were done in both academia and telecommunication industries in characterizing the performance of the SIP proxy server with the latest technological advancement in server architectures.

In order to specifically model the SIP proxy server, it is necessary to understand the design and functions of a SIP proxy server. Since we intend to compare the IP telephony services with PSTN services, we briefly addressed the PSTN functions and call setup as well. PSTN provided several services with the help of a well-known service layer called Advanced Intelligent Networks (AIN) (Lazar et al. 1992). PSTN is a centralized network switch also known as Central Office [CO]. Signalling System # 7 (SS7) is a widely known protocol used to route the signaling messages with media streams between the central offices (Lin 1996). Recently, IP telephony has become widely accepted as a next-generation technology for the communications infrastructure (Gurabani 2004). Internet telephony consistently provides the real-time transportation of voice, and video based application services (Chatterjee et al. 2005). SIP is ASCII-based, resembling Hypertext Transfer Protocol (HTTP), and reuses existing IP protocols such as Domain Name Service (DNS), Real Time Protocol (RTP), Session Description Protocol (SDP) etc. to provide media setup and teardown (Rosenberg et al. 2002). The SIP proxy server is a call control software package that enables service providers to build scalable, reliable Voice over IP networks today and it provides a full array of call routing capabilities to maximize network performance in both small and large packet voice networks. The SIP proxy server can perform a digest authentication of SIP Register, invite requests, and can encrypt SIP requests and responses using Transport Layer Security (TLS). For secure communication, user authentication, confidentiality and integrity of signaling messages and SIP session establishment are essential. TLS is used for the secured SIP signaling and Secure Real Time Protocol (SRTP) for secured SIP session establishment.

## 1.1 Contributions

The key contributions of the book are five-fold and are as follows:

- First, we examined the $M/M/1$ queuing network model proposed in Gurbani et al. (2005), obtained the analytical results and compared them with experiments performed on a real SIP proxy server, and report on those results here, which shows that the model does not have sufficient predictive value. Then, we attempted to use an $M/D/1$ queuing network model based on our understanding of the internals of the SIP proxy software. Based on our data, we eventually proposed a simpler alternative analytical model using a small number of queues, which nevertheless predicted the proxy server performance well.
- Second, in the course of this work, we became aware that the typical software realization of the SIP proxy server was undergoing a change. In keeping with the current trends in SIP proxy server architecture, we eventually proposed a surprisingly simple SIP proxy server model based on a single $M/M/c$ queue, motivated by our understanding of the software system that realizes SIP proxy servers in practice. We studied it analytically and experimentally, finding our model to have good predictive value regardless of its simplicity.
- Third, based of the determinations made in the second part of our research, we expanded our research and focused on the performance and scalability of the SIP proxy server with a single proxy server in the SIP network under Local Area Network (LAN) environment. We measured key performance metrics of the SIP proxy server such as average response time, mean number of calls in the system, queue size, memory and cpu utilization.
- Fourth, to understand better on the impact of the SIP proxy server performance and scale when we used different CHTs during the experiments, we did an empirical study with different CHT data and derived inferences that the impact is significant and deserve further detailed investigation. Based on the established $M/M/c$ based proxy server model, we expanded the research to study the performance and scalability of the proxy server using a IETF standard trapezoidal SIP network with two proxy servers in a local and wide area networks. We conducted a comparative study the predicted and experimental results and also studied the performance impact of CHT when the calls are made through LAN and WAN.
- Finally, the fifth part of our research, with the established $M/M/c$ based proxy server model in a LAN environment, studied the performance impacts on the proxy server when secured and non-secured transport protocols are used to establish the SIP call. During this work, we compared the performance of User Datagram Protocol (UDP), Transmission Control Protocol (TCP) in a non-secured transport protocol, TLS-authentication and TLS-Encryption as part of secured transport protocols in a SIP network.

A problem with studying the performance characteristics of SIP interactions is that there is no simple method to compare theoretical predictions with actual data, because there are no standard simulation tools for SIP proxy servers or other parts of the SIP system; nor is it possible to have simulations significantly simpler than the system itself because the system is largely software anyway. This may in part explain why previous studies such as Gurbani et al. (2005) have stopped with proposing a model for such servers and studying the behavior of the model, and have made no attempts to validate their models against any real data. We contributed significant effort into designing the testbed topology; installing and configuring required hardware, studying several required tools, and software to validate our research in the actual lab environment. In all of our studies, we gathered actual experimental data by setting up actual SIP proxy systems and software phones in the lab. We see this as one of our major contributions. It may allow simpler simulation models of SIP proxy to be used by later researchers.

At the beginning of the book we wanted to focus on the performance of the proxy server rather than the network connecting it, and so have considered only a single proxy server processing the entire incoming SIP calls generated from SIP endpoints, and assumed no delay due to proxy servers located at distant locations. Detailed the SIP proxy server performance with the current IETF recommended standard trapezoidal SIP architecture with two-proxy server model.

## 1.2  Structure of the Book

In the first chapter, the context of PSTN and VoIP services context is presented, which sets the foundation for this research. Chapter 3 surveys the recent literature related to this research; Chap. 4 describes the emulated $M/M/1$ model with predictions of the model; followed by the experimental Data set 1 results are presented and then proposed an analytical $M/D/1$ model along with the comparative study of the SPS performance; Presented a complete study of current SIP proxy server architecture, key components and functions in Chap. 5; Proposed $M/M/c$ based queuing model with predictions of the model, followed by the experimental data set and then the comparative study of the performance models of the SPS in Chaps. 6 and 7 details the SIP proxy server performance when the calls are made through local and wide area networks with the comparative study of the SPS performance and in Chap. 8, studied the performance impacts on SIP proxy servers when SIP security protocols are used compared to non-secure transport protocols. And finally Chap. 9 addresses the Statistical regression analysis for all the experimental data sets of the SIP proxy server.

# Chapter 2
# PSTN and VoIP Services Context

## 2.1 SS7 and PSTN Services Context

### 2.1.1 PSTN Architecture

During the 1990s, the telecommunication industries provided various PSTN services to the subscribers using an Intelligent Network (IN) called Signaling System Number 7 (SS7). SS7 was considered one of the most reliable and capable signaling protocols. IN is defined as an architectural concept for the operation and provision of services as recommended by ITU-T standard. Workstations installed with software interfaces with the network can provide many advanced services to the subscribers. Advanced Intelligent Network (AIN), the later version of IN, provided several telephone services like 800 service, call forwarding, call waiting, call pickup, follow me diversion, store locator service, and name delivery, which are valuable services to the traditional phone users. All around the world, telecommunication companies deployed the PSTN network for their subscribers to communicate from their homes or offices. Telephone subscribers are normally attached to a central office within their geographic area. These central offices have large-scale computers, generally called switches, built with huge hardware and managed by various software modules. These switches are considered as the brain of the PSTN network. Normally, there are several central offices within a metropolitan area. There are two widely used call types: (1) Local call where the called party is within the geographic area of the central office; (2) Long distance call or tandem call where the traffic outside the central office goes to one or more toll/tandem offices, which contains a tandem switch. To connect and send/receive traffic between the central offices and the tandem switches, trunks are being used. Routing the signaling messages with media streams between the switches over a packet-based network is the main function of SS7. Traditional telephone network has been deployed with advanced services to the telephone subscribers since the introduction

S.V. Subramanian and R. Dutta, *Measuring SIP Proxy Server Performance*,
DOI 10.1007/978-3-319-00990-2_2,
© Springer International Publishing Switzerland 2013

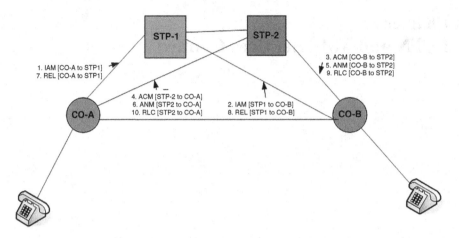

**Fig. 2.1** SS7 basic call setup

of digital telecommunication switches. In the 1980s, the International Telegraph and Telephone Consultative Committee (CCITT) standardized the IN services in Q.1201. PSTN subscribers throughout the world are overwhelmed by the advanced services like call forwarding, call waiting, voice mail, speed dial, call return, and other simple user location services. ITU-U standard defines IN as a conceptual model and the "Framework for the design and description of the IN architecture".

## 2.1.2   PSTN Call Setup

The subscriber attached to originating central office-A (CO-A) initiates the call by dialing the digits on the calling party telephone Fig. 2.1. Central office-A collects the digits, checks for available trunks, builds an Initial Address Message (IAM) and routes it to the destination exchange or central office-B. Central office-B verifies the called number and the availability of the subscriber; the phone rings and then sends an Address Complete Message (ACM) to central office-A. Then the called party answers the phone and the central office-B sends an Answer Message (ANM) back to the originating central office-A. Now there is a voice path established between the calling and called parties. After the conversation between both the parties is complete, called party hangs the phone, then the central office-B builds and sends Release Message (REL) to the originating central office-A. Finally, the originating central office-A acknowledges the REL with Release Complete Message (RLC) and sends it to the destination central office-B (Lazar et al. 1992; Lin 1996).

## 2.2 SIP and IP Telephony Services Context

### 2.2.1 IP Telephony Services

IP telephony is considered as the next-generation solution for the communications infrastructure. To support an IP telephony solution, the enterprise network must be a multi-service network–a network on which voice, video and data can coexist over a single IP-based infrastructure. Internet telephony provides the real-time transportation of voice, and video based application services (Chatterjee et al. 2005). Unlike well-defined PSTN service architecture, IP telephony architecture is still under construction and is not stable. In PSTN, the signaling traffic uses a separate network from the media traffic where in IP telephony both signaling and media traffic are in the same network. In PSTN, the signaling messages flow through the intermediate components for the entire call duration. In IP telephony, the signaling messages are routed through the core intermediate components until the session is established and then the media streams flow directly between the endpoints. Unlike PSTN, there is no centralized service execution platform in IP telephony. In reality the endpoint devices used in the IP telephony such as Personal Computers (PC), laptops, cell phones, and Personal Digital Assistants (PDA) are most powerful than the PSTN endpoint like a simple telephone. In PSTN, the services reside at the core of the network (IN entities), whereas in the IP telephony services reside at the endpoints as shown in Fig. 2.2. The core of the Internet is simplistic in nature and performs the routing services only. Due to lack of centralized service control point, the Internet endpoints can host and perform many IP telephony services. Deploying IP telephony service at the endpoint provides a mixed result in reality. SIP and H323 IP signaling protocols provides service specifications in IP telephony on top of the signaling functions such as call setup, media stream modifications, and terminating the call. Services in IP telephony are created by SIP common Gateway Interface (CGI) (Lennox et al. 2001), Call Processing Language (CPL) (Lennox et al. 2004) or SIP servlets (Sun Microsystems 2003).

### 2.2.2 SIP Notations and Terminologies

SIP is an application layer signaling protocol that can initiate, modify, and terminate media sessions over Internet such as IP telephony calls. SIP packets are text based, transaction based, easy to read, easy to debug, and easy to develop new services more effectively. SIP transports voice, video, and data using User Datagram Protocol (UDP), Transmission Control Protocol (TCP), and Transmission Layer Security (TLS) as a transport layer protocol. SIP provides session establishment, session transfer, session termination, and session participant management (Example: Presence) (Johnston 2003). SIP also provides user location services such as address resolution, name mapping and call redirection. SIP user agents (SIP

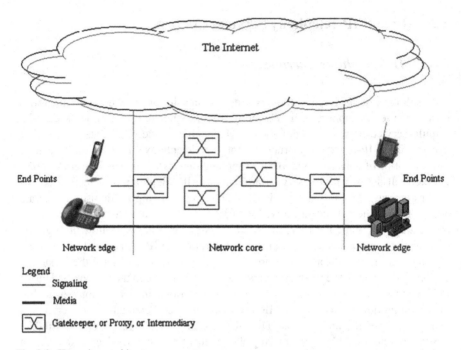

The Internet

End Points                                                                                End Points

Network edge                          Network core                          Network edge

Legend
——— Signaling
▬▬▬ Media
[X] Gatekeeper, or Proxy, or Intermediary

**Fig. 2.2** IP services architecture

UA) are the end user devices such as PCs, PDAs, and cell phones that are used
to create and manage SIP sessions. SIP Registrar server is a network server which
accepts/stores/serves registration requests and may interface with location services
such as Lightweight Directory Access Protocol (LDAP), Common Object Request
Broker Architecture (CORBA) and Open Database Connectivity (ODBC) database
servers (Johnston 2003). SIP addresses are called Uniform Resource Identifier
(URI). URIs is of the user@host format or E164 number@host format. URIs do
not directly refer to the transport address but it are an abstract entity that can
reach the user either directly or indirectly (Rosenberg et al. 2002). The INVITE
request sets up the call. The BYE request terminates the call. The REGISTER
request registers the endpoints with the SIP registrar server. The UPDATE request
updates the original invitation. The ACK message represents reliability and call
acceptance. The OPTIONS message is used for querying the participants about
the media capabilities. The CANCEL message is used for terminating the INVITE
request. There are six different responses in SIP as shown in Fig. 2.3.

- 1xx responses are informational responses also known as provisional responses
  (100 Trying, 180 ringing, 183 Session in progress).
- 2xx responses are successful responses also known as final response (200 OK,
  202 accepted)

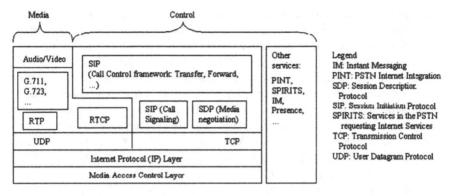

**Fig. 2.3** SIP stack

- 3xx responses are re-directional responses (301 temporarily moved, 302 permanently moved).
- 5xx responses are server failure responses (500-server error).
- 6xx responses are global failure responses (604 not found anywhere) (Lennox 2004).

### 2.2.3 SIP Architecture

There are four key components in SIP.

- SIP User Agents (UA)
  SIP user agents (UA) are the end user devices such as PCs, PDAs, and cell phones that are used to create and manage SIP sessions. User Agent Client (UAC) sends the SIP requests. User Agent Server (UAS) listens to the SIP requests and sends response back to UAC. Back-to-back User Agent (B2BUA) is the concatenation of UAC and UAS.
- SIP Registrar Server
  SIP Registrar server is a network server which accepts/stores/serves registration requests and may interface with location services such as LDAP, CORBA and ODBC database servers.
- SIP Proxy Server
  SIP proxy server accepts the session request sent by SIP UA, and queries the SIP registrar server to obtain the recipient UA's addressing and routing information (Malas 2009). SIP proxy server then forwards the session invitation directly to the recipient SIP UA if it is in the same domain or to another proxy server if the UA resides in a different domain. A SIP proxy server initiates requests on behalf of and receives requests from a client. More formally, a SIP proxy server is an intermediate entity that acts as both a server and a client for

the purpose of creating requests on behalf of other clients. A SIP proxy server primarily handles routing, and ensuring that a request is sent to another entity closer to the targeted user. SIP proxy servers are also useful for enforcing policy (for example, ensuring that a user is allowed to make a call). SIP proxy servers provide functions such as authentication, authorization, network access control, routing, reliable request retransmission, and security. They are often co-located with redirect or registrar servers (Johnston 2003). SIP proxy server can use any physical-layer interface in the server that supports IP. Users in a SIP network are identified by unique SIP addresses. A SIP address is similar to an e-mail address and is in the form sip:userID@gateway.com (Rosenberg et al. 2002). The user ID can be either a username or an E.164 address. Users register with a registrar server/SIP proxy server using their assigned SIP addresses. The registrar server provides this information to the location server upon request. A user initiates a session by sending a SIP request to a SIP server (either a SIP proxy server or a redirect server). The request includes the Address of Record (AOR) of the called party in the Request URI.

• SIP Redirect Server
  SIP redirect server allows the SIP proxy server to direct SIP requests to external domains. SIP proxy server and SIP redirect server can reside in the same hardware (Gurabani 2004).

**SIP Generic Call model:** A UAC can directly contact a UAS if it knows the location of the UAS and does not want any special services from the network. However, a UAC typically initiates a call through a proxy server and relies on the proxy server to locate the desired UAS and obtain any special services from the network. The SIP messaging path from UAC to UAS can involve multiple proxy servers, and in such scenarios SPS interfaces at a peer level with other proxy servers. SIP requests can be sent with any reliable or unreliable protocol. SPS supports the use of User Datagram Protocol (UDP), Transmission Control Protocol (TCP), and Transport Layer Security (TLS) for sending and receiving SIP requests and responses (Rosenberg et al. 2002). The media stream established between two SIP endpoints is depicted in Fig. 2.4.

## 2.2.4  SIP Session Setup Within Same Domain

After the SIP UA Client (UAC) [PDA, PCs, and laptop] and SIP UA Server (UAS) [laptop, PDA, PCs] devices are powered on, their IP addresses and the availability are registered with the SIP registrar server, as shown in Fig. 2.5. When a UAC intends to establish a multimedia session with the UAS, it sends an invitation to the proxy server to connect to UAS. The proxy server queries and receives the UAS's IP address and other routing information from the SIP registrar server. The SIP proxy server relays UAC's invitation to communicate with UAS. If UAC's invitation is acceptable then UAS informs the SIP proxy server that it is ready to

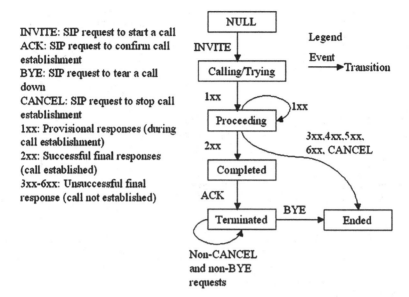

INVITE: SIP request to start a call
ACK: SIP request to confirm call
establishment
BYE: SIP request to tear a call
down
CANCEL: SIP request to stop call
establishment
1xx: Provisional responses (during
call establishment)
2xx: Successful final responses
(call established)
3xx-6xx: Unsuccessful final
response (call not established)

**Fig. 2.4**  SIP call setup and tear down state diagram

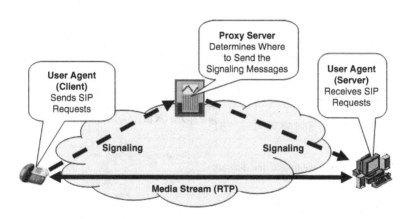

**Fig. 2.5**  SIP signaling

receive packets. The proxy server communicates the acceptance of the invitation
to UAC and the UAC acknowledges (ACK) the acceptance. Finally the multimedia
session between UAC and UAS is established. To terminate the session either UAC
or UAS sends a BYE request. Figure 2.6 details the exchange of SIP requests and
response messages between UAC and UAS through a SIP proxy server, since this
example SIP session is within the same domain. SIP session and SIP call are used
interchangeably throughout this book (Fig. 2.7).

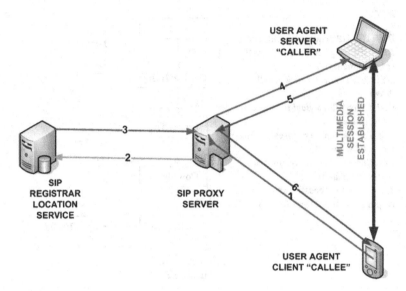

**Fig. 2.6** SIP call within the same domain

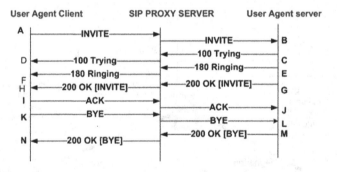

**Fig. 2.7** SIP basic message transaction diagram

## 2.2.5  SIP Security

The deployment and delivery of SIP value-added services in the public network carries security issues that end users and service providers must understand. Service providers carry the biggest responsibility, as they have to offer a secure and reliable service to the end user. Service providers must show that this value-added service does not compromise existing security and that the end user's public presence is protected and managed. Deploying SIP services in the network exposes its core and edge entities to security threats such as network Denial of Service (DoS) attacks, SIP level floods and other SIP vulnerabilities (Cha et al. 2007). Floods of SIP packets can overwhelm servers, bandwidth resources can be consumed and the quality of voice and video over IP can be degraded. If any of these activities are left unchecked, SIP server crashes may result, hindering or even completely paralyzing a business's SIP functionality. Also by nature Internet based applications such as VoIP are vulnerable to several security attacks. It is imperative to provide security to SIP signaling and media stream after establishing the SIP session. TLS and SRTP provides the total security for SIP signaling and the secured media session between two SIP User Agents (SIP UAs) (Dierks and Allen 1999).

# Chapter 3
# Related Work

Both academia and telecommunication industries did several studies and published many literatures on the performance and reliability of PSTN based SS7 signaling. Authors in Wang (1991) studied the performance analysis of link sets on the CCS7 network based on SS7 signaling with complete derivation on the analytical queuing and delay models. Several performance models are based on SS7 traffic load matrix and traffic characteristics, network configuration and network protocols (Kohn et al. 1994). The same paper addresses the reliability for SS7 networks based on availability, dependability and robustness. In Chung et al. (1999), the team calculated the mean end-to-end delay of Common Channel Signaling Networks [CCSN] and used call arrival rate and failure rates. Total response time, network transmission time, and SCP processing are the key performance parameters considered by the work done in Atoui (1991).

Wu (1994) presented the performance modeling, analysis and simulation of SIP-T (Session Initiation Protocol for Telephones) signaling system in carrier class Voice over IP (VoIP) network. The authors used queuing theory and derived the analytical model for throughput efficiency. The SIP-T signaling system defined in Internet Engineering Task Force (IETF) draft is a mechanism that uses Session Initiation Protocol (SIP) to facilitate the interconnection of PSTN with carrier class VoIP network. SIP-T is a scalable, flexible and interpretable signaling protocol with PSTN and also provides call control function of the Media Gateway Controller (MGC) to setup, tear down and manage VoIP calls in the carrier class VoIP network. The authors analyzed the appropriate queuing size and came up with the mathematical model for finding the queuing delay, and the variance of queuing delay of SIP-T signaling system that are the major evaluation parameters for improving Quality of Service (QoS) and the system parameters of MGC in carrier class VoIP network. The theoretical model is based on the $M/G/1$ queuing model, i.e. they assumed Poisson distribution for the arrival process of the SIP-T messages and general service time of the SIP-T message. They also used $M/M/1$ queuing model to observe the number of SIP-T messages in the system. They came up with the derivation for the queuing

S.V. Subramanian and R. Dutta, *Measuring SIP Proxy Server Performance*, DOI 10.1007/978-3-319-00990-2_3,

size using the Embedded Markov chain and semi Markov chain. Authors also compared their theoretical performance model with the actual results to show its very consistencies.

Average number of SIP-T messages M in the system is modeled with the queuing size using the Embedded Markov chain ($M/M/1$ queuing model)

$$M = \frac{(1-a)(r^2h_0^2 + 2rh_0) + (r^3h_0h_1^2)}{2(1-a)(rh_0 + 1 - a)} where\ k = 1,2,3,\ldots N, r = 0,1,2., \quad (3.1)$$

$h_1$ is the mean service time of the SIP-T message, $h_0$ is the mean service time of the fill-in message when k=1 and r = 0. $a = rh_1$ is the mean traffic intensity of the SIP-T message. In Eq. 3.1, the observation time is a constraint to the departure instant of a message.

Using the arbitrary instant of a message and the semi Markov chain, the average number of SIP-T messages N in the system is derived as:

$$N = a + \frac{rh_0}{2} + \frac{r^2h_1^2}{2(1-a)} \quad (3.2)$$

Most of the research in IMS/VoIP related to SIP is focused on engineering principles, protocol definitions, enhancement and other improvements. Very little research work is done in the area of performance modeling of SIP proxy servers. Recently, lot of IP telephony industries are focusing on various SIP and SIP proxy server performance metrics (Malas 2009).

Schulzrinne et al. (2002) approximately dimensioned the SIP-based IP telephony systems and found that the SIP registration rate can reach thousands of requests per second. They proposed an initial simple set of performance benchmarks for SIP proxy, Redirect and Register servers. This research builds in part on SIPstone to provide analytical and real world solution for the better understanding of SIP proxy server performance.

Rajagopal and Devetsikiotis (2006) analyzed and proposed the IP Multimedia Subsystem (IMS) network based on the SIP signaling delay and predicted performance trends of the network, which allowed them to choose parameter values optimally. The proposed models were based on the queuing models for the IMS network and characterization of the SIP server workload, and on a methodology for the design of such networks for optimal performance in a real time SIP traffic. They analyzed the IMS performance as an end-to-end delay point of view. They defined a utility function in their model for the IMS network to optimize the service rate. They assumed the mean service rate to be greater than or equal to mean arrival rate to model the fast connection setups. The utility function of an IMS network is modeled as $U = V(\lambda) - C - P$ where $U$ is the utility, $V(\lambda)$ is the revenue earned by serving $\lambda$ connections, $C$ is the cost compensation for maintaining $K$ servers in the network and $P$ is the penalty function representing revenue losses due to lost connections.

The optimal utility function is:

$$\hat{U} = max_\rho \left[ V - K(a\rho - b\rho^2) - \beta T - \frac{KT}{\rho - \lambda} \right] \tag{3.3}$$

Bozinovski et al. (2002) presented a performance evaluation of a novel state-sharing algorithm for providing reliable SIP services. SIPs built in state machines state information being used instead of transport protocol states. The state-sharing algorithm saves and retrieves the SIP state information using queues. Failure detection mechanism uses the SIP retransmission timers to determine the actual failure of calls. The fail over mechanism defines the primary proxy and backup proxies and dynamically changing the destination proxy information using the round robin algorithm. The authors considered server failure and repair time distributions; call arrival and call duration distributions are their most important performance parameters.

Cursio and Mundan (2002) analyzed call setup time in the SIP based video telephony. Authors used a 3G-network emulator to measure post-dialing delay, answer-signal delay and call release delay. The results are compared to local, national and international Intranet LAN calls. They also studied the effect of SIP calls over channels with restricted bandwidth, typical of mobile network signaling bearers. They considered the SIP based video IP telephony and the transaction based delay; meaning delay during the SIP transaction and session establishment. They came up with the SIP call setup delay over 3G networks and also did a comparison with Internet LAN results.

Fathi et al. (2004) focused on SIP protocol and evaluated its session setup delay with various underlying transport protocols [TCP, UDP, Radio Link Protocol(RLP)] as a function of FER. They defined the Call setup delay as the message that can travel between the UAC and UAS. Provisional responses such as 100 trying and 180 ringing messages are not part of the delay. They did not consider the processing delays and queuing delays through different proxy servers. They also came up with the following results: For a 9.6 Kbps channel, the SIP session setup time can be up to 84.5 s with UDP and 105 s with TCP when the FER is up to 10 %. The use of RLP puts the session setup time to 2.4 s under UDP and 3.5 s under TCP for the same FER and the same channel bandwidth.

Eyers and Schulzrinne (2000) outlines a simulation study of the Internet Telephony call setup delay, based on UDP delay/loss traces. The focus is on the signaling transport delay, and the variations arising from packet loss and associated transmissions. Results indicate during high error periods, H323 call setup delay significantly exceeds that of SIP. INVITE over UDP retransmitted until first provisional response (1xx) arrives, at 0.5, 1, 2, 4, 8, and 16 s. Final response (2xx or higher) are retransmitted until ACK arrives, with the interval capped at 4 s. Final responses are retransmitted when retransmitted INVITE arrives. Retransmissions are limited to 7 packets or 32 s. Other methods (BYE, OPTIONS, and CANCEL) do not retransmit responses a cap timeout is 4 s.

Bozinovski et al. (2002) evaluated the performance of SIP based session setup over the S-UMTS, taking into account the large propagation delay over the satellite as well as the impact of the UMTS radio interface. It also addresses the issue of incorporating a link layer retransmission based on the Radio Link Control acknowledgement mode (RLC-AM) mechanisms and examines its implications on the SIP session establishment performance.

Towsley et al. (2001) developed an inference models based on the finite capacity single server queues for estimating the buffer size and also studied about the intensity of the traffic. There came up with the mathematical derivation for their model to find out the server utilization.

$$U = \rho(1 - P_L) \tag{3.4}$$

where Loss Probability

$$P_L = \frac{(1-\rho)\rho^k}{1-\rho^{k+1}} \tag{3.5}$$

where k represents number of customers in the queue.

Average Response time:

$$R = \frac{1}{\mu(1-\rho)} - \left(\frac{k}{\mu}\right)\left(\frac{\rho^k}{1-\rho^k}\right) \tag{3.6}$$

where traffic intensity $\rho = \frac{\lambda}{\mu}$ and $\lambda, \mu$ represents arrival and service rates.

Gurbani et al. (2005) came up with an analytical SIP based performance and reliability model, in which they primarily considered the mean response time and the mean number of calls in the system. Mean response time of a proxy server is the difference between the times it takes for an INVITE sent from a UAC to reach the SIP proxy server until the final response is sent by the SIP proxy server to the UAC. Mean number of calls is defined as the mean number of sessions that are currently in the system. They modeled a SIP proxy server as an open feed forward queuing network. The INVITE requests are forwarded upstream to the SIP proxy server from the UAC. The SIP proxy server is modeled as queuing model with six queues. Their model is based on $M/M/1$ queue. Mean number of calls N (random variable) in the system at study state is given by:

$$N = \sum_{k=1}^{J} \frac{\rho_k}{(1-\rho_k)} \tag{3.7}$$

where

$$\rho_k = \frac{\lambda_k}{\mu_k}, \lambda_1 = \lambda, \lambda_j = \sum_{k=1}^{j-1}(\lambda_k Q[k,j]) \tag{3.8}$$

for $1 < j \le J$ and $J = 6$ is the number of stations in the queuing model. $Q$ is the one step probability matrix corresponding to the queuing model, that is, $Q[i, j]$ the probability that a job departing station $i$ goes to station $j$. The mean response time for calls is by Little's law $R = \frac{N}{\lambda}$. The authors assumed the service rate is fixed at $0.5 \, \text{ms}^{-1}$ and the arrival rate at $0.3 \, \text{ms}^{-1}$.

Guillet et al. (2008) proposed a mutual authentication mechanism and also discussed several possible security attacks on SIP endpoints and SIP proxy server. They proposed a mutual authentication mechanism within HTTP Digest, provides meaning and semantic to some of the parameters' values generated by the participating end-points during SIP session establishment, particularly the nonce values. They claimed that their approach helps in reducing DoS (Denial of Service) attacks, detects server identity spoofing and ensures basic mutual authentication with comparison to HTTP digest.

Cha et al. (2007) demonstrated the effect of the security protocol on the performance of SIP by comparing the call setup delays that occur with various security protocols. They simulated the various combinations of three security protocols and three transport – layer protocols recommended for SIP. Their result indicates the UDP with any combination of security protocols performs much better than the combination of TCP. They also concluded TLS over Stream Control Transmission Protocol (SCTP) may have more impact on performance on average.

Yanik et al. (2008) studied the SIP proxy server performance and authentication cost for four different SIP proxy server software. SIP proxy server is based on several factors such as SIP stack parsing, processing time and queuing time, they compared the performance of four different well-known SIP proxy server software.

Kim et al. (2008) analyzed and implemented security protocols for SIP based VoIP and studied the performance of their security protocol implementation. Their implementation is partly based on TLS. Implemented new security mechanism to protect the SIP signaling and RTP session. Applied their implementation to VoIP hardware such as SIP phones, SIP proxy and evaluated the performance about call setup delay and voice quality.

Ram et al. (2008) found while comparing the performance of the OpenSER SIP proxy server software (IPTEL 2002) on UDP verses TCP, and concluded the software implementation causing the TCP to perform poorly. They found principal reason for the poor performance of TCP is the architecture of the SIP proxy server that supports connection-less protocol such as UDP than a connection-oriented protocol such as TCP. They also addressed the issue with the software and proved the performance of TCP is better than before.

# Chapter 4
# Performance Measurements of $M/M/1$ and $M/D/1$ Based SPS

At the beginning of our research, we identified that there are very limited research is being done on SIP proxy server performance and VoIP in general. In a very fast phase, due to the growth of Internet based technologies, IP telephony quickly became a direct replacement for PSTN. Lack of any SIP proxy server performance study is mainly because of the complexity involved in simulating or emulating the actual SIP proxy server performance models to accurately study the performance characteristics. We were motivated, excited and equally challenged to work in this area of research. We carefully evaluated our lab resources at CISCO "LASER" lab and leveraged that in our research to perform experiments with the actual proxy servers and collected our data.

## 4.1  Motivation

Authors in Gurbani et al. (2005) have stopped with proposing a performance and reliability model for the SIP proxy server and have made no attempts to validate their model against any real data. This part of the research is to validate their model by conducting experiments in the lab, capture the performance data of the SIP proxy server, parse the data, analyze the data and compare that with the analytical results. It's extremely difficult to study the performance characteristics of SIP proxy servers using any known simple simulation and obtain accurate performance data. Designed the appropriate experiment setup with state of the art equipment, installed, configured and provisioned them in the lab, performed several experiments with SPS software, and used several 3rd party tools to conduct the experiments. In this part of our research, we validated the Gurbani et al. (2005) model and then proposed an alternative simpler analytical model with $M/D/1$ queues.

S.V. Subramanian and R. Dutta, *Measuring SIP Proxy Server Performance*,
DOI 10.1007/978-3-319-00990-2_4,
© Springer International Publishing Switzerland 2013

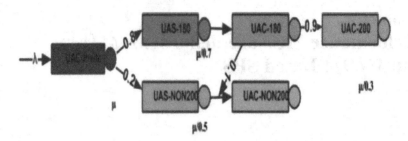

**Fig. 4.1** SIP proxy server $M/M/1$ queuing model proposed in Gurbani et al. (2005)

## 4.2   $M/M/1$ **Model Emulation**

To assess the usefulness of the queuing model proposed by Gurbani et al. (2005), first emulated this model to validate their results with analytical and experimental results (the authors of Gurbani et al. (2005) only provided the analytical model and its consequences but did not provide a comparison with real, emulated, or simulated performance data). In this work, the SIP proxy server is modeled as open feed-forward queuing network as shown in Fig. 4.1, in which incoming INVITE message from UAC as arrival calls and there are sequence of 6 $M/M/1$ based queuing stations that corresponds to each SIP message in as shown in Fig. 2.6. All the incoming SIP packets are processed by executing the same SIP proxy server software module for each session to establish and torn down. To validate the analytical solution, several sets of experiments were conducted with the real SIP proxy server that is built with the emulated $M/M/1$ queuing model based software, collected the data and compared that with the analytical results. SIP proxy server model and assumptions of Gurbani et al. (2005) were;

- Service time $1/\mu$ is the mean service time to process the INVITE request at the SIP proxy server and that service rate is fixed at $0.5\,\text{ms}^{-1}$.
- Since UAS does not parse the SIP packets, the computation will be less, hence the service times are assumed as $0.7/\mu$ for sending the 180 followed by 200 with $0.3/\mu$ or non 200 response with $0.5/\mu$. Analytical and experimental results are shown in Tables 4.1, 4.2 and Figs. 4.7–4.9.
- Only 80 % of the INVITE messages will be successful in getting the 180 response and 90 % of that 180 responses will get the 200 response. Remaining 20 % of the INVITE message will get a non-200 response and 10 % of the 180 responses will receive a non-200 response.

## 4.3   **Hardware and Lab Setup**

This part of research, experiments were conducted with a SER open source (IPTEL 2002) SPS software, to validate the performance model presented by authors

```
<scenario name="Basic UAC">

 <send retrans="500">
   <![CDATA[
     INVITE sip:[service]@[remote_ip]:[remote_port] SIP/2.0
     Via: SIP/2.0/[transport] [local_ip]:[local_port];branch=[branch]
     From: sipp <sip:sipp@[local_ip]:[local_port]>;tag=[call_number]
     To: sut <sip:[service]@[remote_ip]:[remote_port]>
     Call-ID: [call_id]
     CSeq: 1 INVITE
     Contact: sip:sipp@[local_ip]:[local_port]
     Max-Forwards: 70
     Subject: Performance Test
     Content-Type: application/sdp
     Content-Length: [len]
     v=0
     o=user1 53655765 2353687637 IN IP[local_ip_type] [local_ip]
     s=-
     c=IN IP[media_ip_type] [media_ip]
     t=0 0
     m=audio [media_port] RTP/AVP 0
     a=rtpmap:0 PCMU/8000
   ]]>
 </send>
 <recv response="100"
       optional="true">
 </recv>
 <recv response="180" optional="true">
 </recv>
 <recv response="200" rtd="true">
 </recv>
 <send>
   <![CDATA[
     ACK sip:[service]@[remote_ip]:[remote_port] SIP/2.0
     Via: SIP/2.0/[transport] [local_ip]:[local_port];branch=[branch]
     From: sipp <sip:sipp@[local_ip]:[local_port]>;tag=[call_number]
     To: sut <sip:[service]@[remote_ip]:[remote_port]>[peer_tag_param]
     Call-ID: [call_id]
     CSeq: 1 ACK
     Contact: sip:sipp@[local_ip]:[local_port]
     Max-Forwards: 70
     Subject: Performance Test
     Content-Length: 0
   ]]>
 </send>
 <pause/>
 <send retrans="500">
   <![CDATA[
     BYE sip:[service]@[remote_ip]:[remote_port] SIP/2.0
     Via: SIP/2.0/[transport] [local_ip]:[local_port];branch=[branch]
     From: sipp <sip:sipp@[local_ip]:[local_port]>;tag=[call_number]
     To: sut <sip:[service]@[remote_ip]:[remote_port]>[peer_tag_param]
     Call-ID: [call_id]
     CSeq: 2 BYE
     Contact: sip:sipp@[local_ip]:[local_port]
     Max-Forwards: 70
     Subject: Performance Test
     Content-Length: 0
   ]]>
 </send>
 <recv response="200" crlf="true">
 </recv>
 <ResponseTimeRepartition value="10, 20, 30, 40, 50, 100, 150, 200"/>
 <CallLengthRepartition value="10, 50, 100, 500, 1000, 5000, 10000"/>

</scenario>
```

**Fig. 4.2** UAC SIPp sample XML code flow chart

```
<scenario name="Basic UAS responder">

 <recv request="INVITE" crlf="true">
 </recv>
 <send>
   <![CDATA[
     SIP/2.0 180 Ringing
     [last_Via:]
     [last_From:]
     [last_To:];tag=[call_number]
     [last_Call-ID:]
     [last_CSeq:]
     Contact: <sip:[local_ip]:[local_port];transport=[transport]>
     Content-Length: 0
   ]]>
 </send>
 <send retrans="500">
   <![CDATA[
     SIP/2.0 200 OK
     [last_Via:]
     [last_From:]
     [last_To:];tag=[call_number]
     [last_Call-ID:]
     [last_CSeq:]
     Contact: <sip:[local_ip]:[local_port];transport=[transport]>
     Content-Type: application/sdp
     Content-Length: [len]
     v=0
     o=user1 53655765 2353687637 IN IP[local_ip_type] [local_ip]
     s=-
     c=IN IP[media_ip_type] [media_ip]
     t=0 0
     m=audio [media_port] RTP/AVP 0
     a=rtpmap:0 PCMU/8000
   ]]>
 </send>
 <recv request="ACK"
       optional="true"
       rtd="true"
       crlf="true">
 </recv>
 <recv request="BYE">
 </recv>
 <send>
   <![CDATA[
     SIP/2.0 200 OK
     [last_Via:]
     [last_From:]
     [last_To:]
     [last_Call-ID:]
     [last_CSeq:]
     Contact: <sip:[local_ip]:[local_port];transport=[transport]>
     Content-Length: 0
   ]]>
 </send>
 <pause milliseconds="4000"/>
 <ResponseTimeRepartition value="10, 20, 30, 40, 50, 100, 150, 20/>
 <CallLengthRepartition value="10, 50, 100, 500, 1000, 5000, 1000/>

</scenario>
```

**Fig. 4.3** UAS SIPp sample XML code flow chart

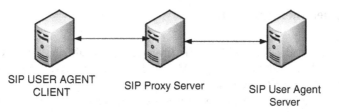

SIP USER AGENT
CLIENT

SIP Proxy Server

SIP User Agent
Server

**Fig. 4.4** Experiment setup

**Table 4.1** Experimental data
set 1 results with SER SPS

| λ (packets/ms) | 0.20 | 0.30 | 0.4 | 0.45 |
|---|---|---|---|---|
| L | 0.82 | 1.49 | 3.28 | 5.68 |
| W in ms | 5.8 | 7.24 | 9.98 | 13.47 |
| $\rho$ in percent | 28 | 46 | 64 | 75 |

**Table 4.2** $M/M/1$
performance model predicted
results

| λ (packets/ms) | 0.20 | 0.30 | 0.4 | 0.45 |
|---|---|---|---|---|
| L | 1.31 | 2.69 | 5.88 | 11.33 |
| W in ms | 7.63 | 9.86 | 15.87 | 26.47 |
| $\rho$ in percent | 40 | 60 | 80 | 89 |

in Gurbani et al. (2005). Configured all the networking elements within a local
Linux based DNS server to avoid any external interference, that can affect the
performance data of the SIP proxy server. All the SIP endpoints are assigned with
an IP address wired to an ethernet switch and a router. Used a most common
configuration of one User Agent Client (UAC) to send the SIP requests, one or
two proxy servers to process the SIP messages, and one User Agent Server (UAS)
to listen to the incoming SIP messages and send responses back to either the SIP
proxy server or to UAC by providing appropriate input variables to the UAC and
UAS xml code as shown in Figs. 4.2 and 4.3. HP 7825H2 servers with the Intel dual
core Pentium D 3.4 GHz processor and 2 GB PC-2-4200 memory are the hardware
loaded with SIP Express Router (SER) version 0.8.14 software (IPTEL 2002),
because SIP Express Router (SER) is a high-performance, easily configurable. SIP-
P tool (Invent 2003) on the server side with the listening port set to 5060 is started
first during the test. Then another instant of SIP-p tool instance started from the UAC
side which can make SIP calls with different call rates. For heavier traffic, JMeter
traffic generator tool (Apache Jmeter 2006) is replaced by SIP-P tool instances by
generate SIP calls from UAC. Performance Monitor (*Perfmon*) tool (XR Perfomon
tools 2006) ran from another HP 7825H2 servers to monitor the traffic to and from
the SIP proxy server. SIP proxy server software that can process all the SIP messages
are instrumented with all the *Perfmon* counters to collect all the performance related
data from the proxy server.

## 4.4    Experiment Procedure

SIP calls are made using SIP-p tool (Invent 2003) or JMeter traffic Generator tool (Apache Jmeter 2006) from the UAC to the SIP proxy server; INVITE messages are processed by the SIP proxy server and then sent to the destination UAS as shown in Fig. 4.4. Responses such as provisional responses (100, 180) and final response (200) are sent to the SIP proxy server and then SIP proxy server forwards that to the UAC. In case of higher call volumes, configured the GUI based JMeter traffic generator tool with appropriate variables like IP addresses of UAC and UAS, IP address of the proxy server, call rate, and others, running on two different machines acting as a UAC traffic generator and UAS traffic generator to generate SIP calls at the rate of 50, 100, 150, 200, 250, 300, 350 and 400 calls per second to the SIP proxy server, which is under test. During the experiment data are collected for 1,000, 2,500 and 5,000 SIP calls. During this time, several statistical data points such as mean response time, queue length, mean number of jobs, server utilization and memory consumption of the SIP proxy server are collected to find any performance impacts on the proxy server. All the SIP call flows can be monitored through *Wireshark - protocol analyzer* tool (Wireshark 2008) for any packet loss and also to validate the accuracy of the statistical data collected by the *perfmon* tool. Refer to Table 4.1 and Figs. 4.7–4.9 for the experimental *Data Set 1* results. Also, verified the results collected using SIPp and JMeter tool for the same call rate and found they are identical or very close.

## 4.5    Predictions from $M/M/1$ Network Model

Queuing Tool Pack 4.0 queuing simulation tool (Ingolfsson and Gallop 2003) is a Microsoft Excel 2007 plug-in (QTP.exe and QTP.dll files) used to calculate the analytical data. The plug-ins can be downloaded into Microsoft office library directory. All the queuing formula and the corresponding calculations can be accessed through the spreadsheet from the "*Formulas menu*" option, then select the "Queuing ToolPak" under category, and then select the appropriate queuing model. We calculated average response time, mean number of number of calls and server utilization data by providing appropriate input values such as $\lambda$, and $\mu$. $M/M/1$ model analytical data is presented in Table 4.2.

## 4.6    Proposed $M/D/1$ Mathematical Model Formulation

Based on the experimental results we obtained during the $M/M/1$ SIP proxy server model, observed that the effort required by the SIP proxy server to process all the SIP requests and responses are almost identical, since the length of the SIP

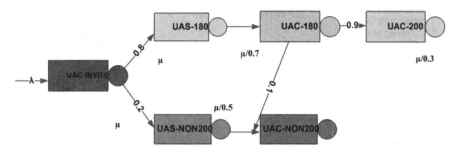

**Fig. 4.5** Proposed SIP proxy server $M/D/1$ queuing model

packet for requests and responses to setup a SIP call is a constant. We proposed a new $M/D/1$ based SIP proxy server model with the arrival rate $\lambda$ and a constant service time $\mu$ for all the SIP request and response messages. Using standard approaches, arrival times are exponentially distributed, service time have no variance and the average queue length is exactly half of $M/M/1$ for a $M/D/1$ queuing model. Figure 4.5 represents the newly proposed SIP proxy server model based on $M/D/1$ queuing model. For each call, all the incoming SIP messages are processed by the SIP proxy server for all the SIP transactions (as shown in Fig. 2.6. There are six or more queuing stations in tandem to process various SIP messages with the deterministic service time. The same SIP proxy server software module is executed for identifying and processing each SIP message based on the unique address for each session until the session is established and tore down.

The average response time, mean number of calls and other performance characteristics for an $M/D/1$ queuing system can be found in the standard derivations and refer to Chap. 11 for the step by step derivation.

$$P(0) = (1 - \rho) \tag{4.1}$$

$$P(1) = (1 - \rho)(e^\rho - 1) \tag{4.2}$$

$$P(N+1) = \left\{ P(N) - [P(0) + P(N)] \times P(N, \rho) - \sum_{k=2}^{N} P(k) \times P(N-k+1, \rho) \right\}$$

$$\frac{1}{P(0, \rho)} \tag{4.3}$$

Since the numerical solution given by the above derivation is proven to be numerically unstable (Abendroth and Killat 2004), we instead use the normalized form of the equation to calculate the mean response time of an $M/D/1$ queue using the famous "Pollaczek - Khinchin (PK)" formula, which can yield a numerically stable solution.

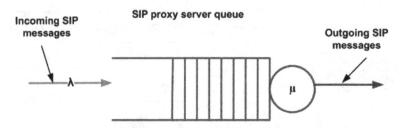

**Fig. 4.6** Proposed SIP proxy server M/D/1 single queue model

$$W = \left[ 1/\mu + \frac{\lambda(1+C_s^2)}{2(\mu)^2(1-\rho)} \right] \qquad (4.4)$$

Using Little's law, we can simply get the mean number in the system as;

$$L = \left[ \rho + \frac{\rho^2(1+C_s^2)}{2(1-\rho)} \right] \qquad (4.5)$$

where $C_s$ is the coefficient of variation. $C_s$ is zero for the deterministic distribution $M/D/1$ queue.

We considered replacing the Gurbani et al. (2005) queuing model with $M/D/1$ with network of queues or simply replacing the entire model by a simple $M/D/1$ with a single queue as shown in Fig. 4.6 for the following reasons:

- The effort required by the SIP proxy server to process all the SIP requests and responses are almost identical or constant service time, whereas they assumed that the INVITE request needed more processing time than other SIP transactions while setting up the SIP call. There is no significant difference in the size of the packets.
- The entire model can be dominated by the constant processing time for initial INVITE requests and other SIP response messages.

## 4.7   Predictions from M/D/1 Models

Downloaded and installed the "Java applet performance measure" tool (Swart and Verschuren 2002) in a PC where Java is installed and enabled. All the queuing formula and the corresponding calculations can be accessed through the GUI. We calculated the average response time and the mean number of calls in the system by providing appropriate input values such as $\lambda$, and $\mu$. Verified the results with another web based $M/D/1$ queuing model simulation tool (Fink 2002). Results are shown in Table 4.3 and Figs. 4.7–4.9

**Table 4.3** M/D/1 performance model predicted results

| λ | 0.05 | 0.10 | 0.15 | 0.20 | 0.25 | 0.30 | 0.35 | 0.40 |
|---|---|---|---|---|---|---|---|---|
| L | 0.11 | 0.23 | 0.36 | 0.53 | 0.75 | 1.1 | 1.6 | 2.5 |
| W in ms | 2.1 | 2.3 | 2.4 | 2.7 | 3 | 3.5 | 4.3 | 6 |

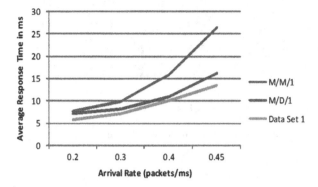

**Fig. 4.7** $M/M/1$ and $M/D/1$ ART verses experimental data set 1

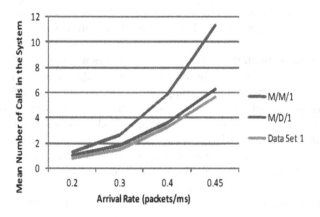

**Fig. 4.8** $M/M/1$ and $M/D/1$ MNC verses experimental data set 1

## 4.8 Comparative Study on Absolute Performance of the SPS

Performance of the SIP proxy servers varies from servers to servers depending on the hardware capacity, CPU, speed and other memory related parameters. The experiments are conducted in a commonly used industrial standard servers like 7825H2 HP hardware and SIP Express Router (SER) version 0.8.14 software (IPTEL 2002) for the SIP proxy server.

- In $M/M/1$ model, we noticed, there are significant differences between the analytical data versus experimental data obtained from the SER real proxy server (see Figs. 4.7 and 4.8), likely due to several assumptions and limitations made

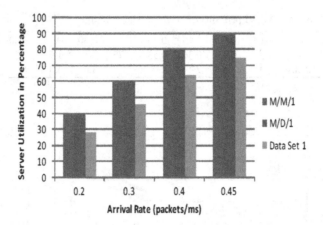

**Fig. 4.9** $M/M/1$ and $M/D/1$ SU verses experimental data set 1

in the analytical model. More importantly, the difference is clearly not just one of scale, i.e. tuning the single $\mu$ parameter in the $M/M/1$ model proposed in previous work would not succeed in matching the experimental data for any value of $\mu$.

- The analytical solution that we obtained from the new proposed $M/D/1$ model with a constant service time for all SIP messages (SIP request and responses), have much better performance compared to the analytical data for the same input parameters and setup used originally for $M/M/1$ based queuing model. The average response times are in the range of 7.63–26.47 ms and mean number of calls are in the range of 1.31–11.33 calls (as shown in Figs. 4.7 and 4.8) in the system in case of $M/D/1$ model compared to average response time of 7.16–16.16 ms and mean number of calls is in the range of 1.03–6.3 calls, incase of $M/D/1$ for the same call arrival rates.

- The average response time to establish a SIP session in $M/D/1$ is less than 50 % compared to $M/M/1$ analytical results and the mean number of calls in the system is also significantly smaller in $M/D/1$ model compare to $M/M/1$ model as shown in Figs. 4.7–4.9.

- The performance of the $M/D/1$ model is better for lower arrival rates and goes down when the call arrival rate increases.

- While comparing the $M/D/1$ network model and $M/D/1$ single queue model with experimental *Data set 1*, we found that the $M/D/1$ single queue model has less significance and not very useful as shown in Figs. 4.10 and 4.11.

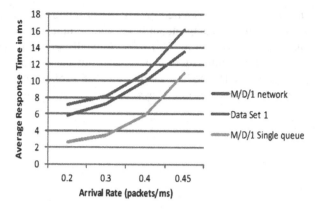

**Fig. 4.10** $M/D/1$ network and $M/D/1$ 1 queue ART verses experimental data set 1

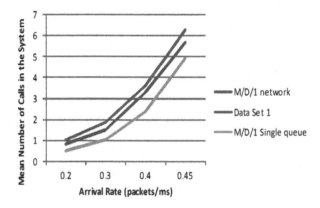

**Fig. 4.11** $M/D/1$ network and $M/D/1$ 1 queue MNC verses experimental data set 1

## 4.9 Concluding Remarks

The performance parameters like average response time and mean number of jobs in the system to setup a SIP based VoIP call is a very fundamental research in IP based telephony that very few researchers are focused on. This study provides a good performance metrics comparison of PSTN services with latest IP based services in the latest IP telephony market place. Based on the experimental results, we determine that the average response time and mean number of jobs, with the single SIP server in a LAN environment, SIP performance data met the ITU-T standards for call setup time. In recent days several advanced programming methodologies and high-speed servers are available, so this performance model can be optimized further, since this queuing model is based on SIP proxy server software. Continue to work on redesigning this queuing model based on the multi-threaded program model that is instead of $M/M/1$ or $M/D/1$ queuing model we intend to focus on $M/M/c$ or $M/D/c$ or the combination of both. Also intend to expand the study by redesigning the performance model with multiple SIP Proxy Servers located in remote locations and factor the network delays.

# Chapter 5
# SPS Software Architecture Study

During the course of this research, we realized that the performance of the proxy server, which is fundamentally a software entity, depends heavily on the software architecture of the system (that is, how the functionality is factored, and apportioned between various threads of execution, and various ways to communicate between functional modules). Decisions regarding this architecture are typically made by the software designers by taking into account the hardware and OS environment that is chosen for the realization, which in turn is driven by business decisions and other factors such as current support and expertise in a particular environment. In recent years, researchers have become aware that the most practically important proxy servers were more and more being based on a simple streamlined execution model, and also were extensively using multi-threading over predecessors. To make the performance study more realistic, it is a reasonable decision to educate on the software architecture of a real proxy server and collect experimental data on it as well. The earlier data set was collected from an open source proxy server implementation running on commodity hardware, as mentioned before. The basic understanding, which changed the modeling effort completely, is that such industry standard servers are increasingly using a single thread of execution to complete all processing needed by a single SIP request completely and sequentially, instead of using the older architecture of multiple concurrently running modules for the various parts of the functionality, coupled by message queues. To utilize the hardware well, concurrency is introduced in these servers by simply spewing a large number of threads, each is identical and carries SIP requests all the way through the processing. We obtained much better understanding of the various processing modules, which we discussed in detail in this chapter.

## 5.1 Address Translation

During address translation, SPS processes the request-URI of an incoming request and returns the list of contacts, each providing a URL for use in the outgoing

S.V. Subramanian and R. Dutta, *Measuring SIP Proxy Server Performance*,
DOI 10.1007/978-3-319-00990-2_5,
© Springer International Publishing Switzerland 2013

request. If number expansion is enabled, SPS applies the global set of expansion rules to the user portion of the relevant URLs for which the host portion is the SPS. For REGISTER messages, this applies to the To, From, Contact, and optionally the authorization headers. For INVITE messages, this applies to the Request URI, from, and optionally the proxy authorization headers (Rosenberg et al. 2002). SPS translation modules, in the order in which SPS calls them, are as follows: (1) Call Forward Unconditional, (2) Registry-the most common use of the translation phase is to perform registry lookups. Users within a domain may REGISTER contact addresses (URIs) for themselves with the proxy. When requests are received in the format of user@localhost, the next hop will be selected from that user's list of active registrations. This is the contact URI (from the header by that name in the REGISTER request), (3) ENUM is a module which converts telephone numbers/extensions within a particular domain into DNS requests for records, and (4) Gatekeeper Transaction Message Protocol (GKTMP): The first module to return one or more contacts completes the translation step, and the remaining modules are not called. For example, if the Registry module returns a contact, then neither the ENUM nor the GKTMP module is called. If none of the translation modules returns a contact, the core proxy module returns a contact based on the incoming Request-URI and that Request-URI is used in the next-hop routing step.

## 5.2   Next-Hop Routing

The next step in SIP request processing is to determine the next-hop route for each contact. Next-hop routing takes each translated request-URI (contact) and locates a set of next-hop SIP entities capable of routing a message to the new request-URI. This step involves two advanced features: SRV (Service) Lookup for Static Route-If the Static Route Next-Hop Port field of a static route is not specified or is zero, SPS performs an SRV lookup on the Static Route Next-Hop field. During location of the next-hop SIP entities, the following occurs: (1) If the host portion of the new request-URI is the address (FQDN-Fully Qualified Domain Name or IP address) of the server itself, next-hop routing is performed by means of the user portion of the request-URI. E.164 routing makes use of this method. (2) If the Static Route Next-Hop port field of a static route is not specified or is 0, SPS tries to do an SRV lookup on the Static Route Next-Hop field. If the lookup succeeds, it uses the algorithm outlined in Gulbrandsen et al. (2000) to select one destination. If the lookup fails, it tries alternate destinations and the proxy server does a simple DNS A lookup on the Static Route Next-Hop field of the static route. This field should contain a name for the A lookup. (3) If the host portion of the new request-URI is not the address (FQDN or IP address) of the server itself, SPS performs domain routing using the host portion of the Request-URI (Mealling and Daniel 2000).

## 5.3 IP Resolution

IP resolution is the conversion of each hop found by means of next-hop routing into an IP address. Standard IP resolution is performed by DNS, Network Information Service (NIS), NIS+, or host file, depending on the IP resolution subsystem configured on the system where SPS is located. If the next hop enters this phase with an explicit IP address and port, the request can be forwarded with any further processing in this phase. Either the next hop is a host name in non-dotted IP format, or the next hop user part represents what looks to the proxy like a telephone number. If the next hop host was taken from a Route header in the request, local (i.e., static) route lookup will be bypassed. In other words, a received request with a Route header containing a host in non-dotted-IP format will skip directly to the resolve via DNS step. Only next hops populated from the received R-URI or from the translation phase, will attempt to route using locally configured static routes. Local/Static Routes If the URI has an explicit 'user=phone' parameter, the proxy will route the request using User Routing. If it has an explicit 'user=IP' parameter, the proxy will route by attempting to resolve the host part of the next hop. User Routing – User routing is the effort by the proxy to route requests by the telephone number found in the user part of the next hop. Since the valid range of telephone numbers overlaps with the username space, in the absence of an explicit 'user=xxx' parameter, there are configuration parameters that help the proxy make this decision. When the proxy attempts to resolve domain routes, it will first check the list of provisioned, i.e., 'static' routes. If no match is found there, DNS SRV (Gulbrandsen et al. 2000) and A record look ups (Mealling and Daniel 2000) will then be performed. If still no match is found, a 404 Not Found response will be returned to the request. Refer to Fig. 5.1 for more details of all the SPS components.

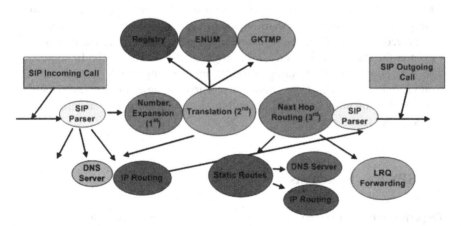

**Fig. 5.1** SIP proxy server software architecture

## 5.4  SPS Components

- **Redirect Servers:** A redirect server is a UAS that generates 3xx responses to requests that it receives, directing the client to contact an alternate set of URIs. It receives requests, strips out the address in the request, check it's address tables for any other addresses that might be mapped to the one in the request, and returns the results of the address mapping to the client for the next one or more hops that a message should take. The client then contacts the next-hop server or UAS directly. Redirect servers are often co-located with proxy or registrar servers.
- **Registry Servers:** A registrar server is a server that accepts REGISTER requests from UACs for registration of their current location. It places the information received in those requests into the location service for the domain that it handles. Registrar servers are often co-located with proxy or redirect servers.
- **Location Services:** A location service is used by a SIP redirect or proxy server to obtain information about a called party's possible locations. It contains a list of bindings of address-of-record keys to zero or more contact addresses. Bindings can be created and removed in many ways; the SIP specification defines a REGISTER method for updating bindings. Location services are often co-located with redirect servers.
- **Registry Database:** The registry database contains location information for registered endpoints. The database is stored in memory-mapped files within shared memory so that information persists between restarts. Registry information is exchanged with all members of a proxy-server farm. The database contains two types of information: dynamic information that is received from endpoints when they register their contact information and static information that is configured by means of the provisioning GUI client. Information pertaining to a single registered user (SIP endpoint) is called a registration. The registry database is thus a collection of registrations.
- **Routing Database:** The routing database contains static route information that the proxy server uses to forward requests toward endpoints that are either not registered with the local registrar server, reside within a different domain, or exist in the PSTN. Static routes are configured based on next-hop IP addresses or next-hop domains. Routing information is configured by means of the provisioning GUI client. As with the registry database, the routing database is stored in memory-mapped files within shared memory so that information persists between restarts.

## 5.5  SPS Key Functions

SPS performs the following steps to deliver messages from endpoint to endpoint:

- Address translation translates an incoming request-URI into an outgoing request-URI.

- Next-hop routing obtains a set of fully qualified domain names (FQDNs) or IP addresses with transport type and port numbers for each of the SIP entities found in the translation step. This step involves the following features and more: (1) SRV lookup for static route-SRV lookup on the Static Route Next-Hop field if the Static Route Next-Hop Port field in a static route is not specified or is zero. (2) Registration, Admission, and Status Protocol (RAS), Location Request (LRQ) message transmission to gatekeeper-Sending the LRQ message to the H.323 gatekeeper to obtain the next-hop gateway transport address.
- IP resolution converts each next hop found in the next-hop route lookup step into an IP address

A proxy server initiates requests on behalf of and receives requests from a client. More formally, a proxy server is an intermediate entity that acts as both a server and a client for the purpose of making requests on behalf of other clients. A proxy server primarily handles routing, ensuring that a request is sent to another entity closer to the targeted user. Proxy servers are also useful for enforcing policy (for example, ensuring that a user is allowed to make a call). A proxy server interprets, and, if necessary, rewrites specific parts of a request message before forwarding it. Proxy servers provide functions such as authentication, authorization, network access control, routing, reliable request retransmission, and security. They are often co-located with redirect or registrar servers. SPS can use any physical-layer interface in the server that supports Internet Protocol (IP).

Users in a SIP network are identified by unique SIP addresses. A SIP address is similar to an e-mail address and is in the form sip:userID@gateway.com. The user ID can be either a username or an E.164 address. Users register with a registrar server/proxy server using their assigned SIP addresses. The registrar server provides this information to the location server upon request. A user initiates a call by sending a SIP request to a SIP server (either a proxy server or a redirect server). The request includes the address or record (AOR) of the called party in the Request URI. Over time, a SIP end user-that is, a subscriber-might move between end systems. The location of the subscriber can be dynamically registered with the SIP server. The location server can use one or more protocols (including finger, RWhois, and LDAP) to locate the subscriber. Because the subscriber can be logged in at more than one station, the server might return more than one address. If the request comes through a SIP proxy server, the proxy server tries each of the returned addresses until it locates the subscriber.

# Chapter 6
# Measurements and Analysis of $M/M/c$ Based SPS Model

With our detailed comparative study of $M/M/1$ model with the experimental results with the open source SIP proxy server software (IPTEL 2002) and also with our proposed analytical $M/D/1$ model, we continued our study more on the internal details of SPS software components, processes and complete software implementations. During the course of this research, we also investigated the limitations of the open source SPS software and replaced that with CISCO SPS software, which is designed, developed and implemented with latest software and hardware technologies for conducting our new set of experiments in the LAN environment. This chapter details the second part of our research in providing some reasonable solutions to the problems identified with our previous study discussed in Chap. 4.

## 6.1 Motivation

A key contribution to this part of research is to obtain a better understanding of the performance of the SIP proxy system and provide an alternate $M/M/c$ based model. Emulated the $M/M/1$ queuing proposed in Gurbani et al. (2005), obtained the analytical results and compared them with experiments performed on a real SIP proxy server using CISCO SPS software for testing instead of IPTEL (2002) software used to collect *Data set 1* discussed in Chap. 4. In the $M/D/1$ model, the SIP Proxy Server software module identifies and processes each SIP message based on the unique address for each session until the session is established and tore down. This creates a variation in queue occupancy at different queuing stations, while processing different SIP transactions, which increases the average response time to setup a session. In this second part of the research, we intend to provide a solution to the problem by proposing a less complex, more predictable, and more realistic SIP proxy server based on $M/M/c$ queuing model, motivated by our understanding of the software system that realizes SIP proxy servers in practice, and studied it analytically and experimentally. Current trends in server architectures makes this

S.V. Subramanian and R. Dutta, *Measuring SIP Proxy Server Performance*,
DOI 10.1007/978-3-319-00990-2_6,
© Springer International Publishing Switzerland 2013

model more suitable than the $M/D/1$ model. In this work, considered only a single proxy server for processing all the incoming SIP calls generated from SIP endpoints in a LAN environment, and hence assumed no delay due to proxy servers located at distant locations.

## 6.2   Proposed $M/M/c$ Based SPS Model

To orient on the latest advancement in software engineering and networking technologies, we selected the CISCO proxy server for this model. SIP proxy server software architecture plays an important role in the parsing and processing of all the SIP packets. Basic software architecture consists of three key modules to process each incoming SIP packet. In the translation module, the Request-URI of an incoming SIP request message is processed, list of contacts returned and each providing a URL to use in the outgoing request. The next-hop routing module determines the next hop routing. Next hop routing takes each translated Request-URI and locates a set of next hop SIP entities capable of routing to the new Request-URI. Then the next module is the IP resolution module, where the conversion of each SIP requests in the next hop routing process into a valid IP address. All the incoming SIP packets must go through all these software modules and it takes some time to process and forward each packet. There is further fine structure in the processing, some of which is shown in Fig. 5.1.

**Key observations are as follows:**  In earlier realizations of the SIP proxy server, it was typical to run each main stage of the processing as a separate standalone concurrent process. In such a realization, each such module is amenable to modeling as a queue, because the coupling between different modules is unsynchronized and through message passing. Such a realization is useful and efficient if the pattern of processing of different packets (i.e. which modules they go through and in what order) varies from packet to packet, and has some statistically (but not deterministically) characterizable patterns. However, it is not any more efficient than a synchronized (function call) processing if almost all packets are expected to go through the same sequence of processing, as is now understood to be the pattern of proxy processing for the INVITE packets which produce most stress on the server. (At best, it is equally efficient; at worst, it is significantly less efficient due to messaging and other overhead.)

The new implementation is based on a single queue with multiple servers, and each packet is run through the entire processing before the next one is picked up. Concurrency can still provide some advantage, but now it is used simply to tune the system capability to the load, not linked to software modules. In other words, the entire processing of each packet shown in Fig. 5.1 is executed using a single thread. This is exactly amenable to modeling by an $M/M/c$ queue, where (c represents number of threads used and it is a constant value) as shown in Fig. 6.1, to process the arrival and departure of SIP message transactions using multi-threaded mechanism by allocating the threads dynamically to process all incoming SIP packets. The value

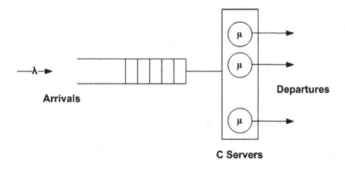

**Fig. 6.1**  SIP proxy server $M/M/c$ queuing model

**Table 6.1**  Experimental data set 2 results

| $\lambda$ (pks/ms) | 0.2 | 0.3 | 0.4 | 0.5 | 0.6 | 0.7 | 0.8 | 0.9 | 1.0 |
|---|---|---|---|---|---|---|---|---|---|
| L | 1.1 | 1.6 | 2.1 | 2.7 | 3.2 | 4.0 | 4.6 | 5.6 | 6.6 |
| W in ms | 5.3 | 5.4 | 5.5 | 5.6 | 5.8 | 6.2 | 6.5 | 7.1 | 7.8 |
| $\rho$ in percent | 11 | 16 | 20 | 25 | 36 | 40 | 47 | 54 | 60 |

**Table 6.2**  $M/M/c$ performance model predicted results

| $\lambda$ (pks/ms) | 0.2 | 0.3 | 0.4 | 0.5 | 0.6 | 0.7 | 0.8 | 0.9 | 1.0 |
|---|---|---|---|---|---|---|---|---|---|
| L | 1.2 | 1.8 | 2.5 | 3.1 | 3.9 | 4.7 | 5.7 | 7.0 | 8.7 |
| W in ms | 6.0 | 6.1 | 6.2 | 6.3 | 6.5 | 6.8 | 7.2 | 7.8 | 8.7 |
| $\rho$ in percent | 13 | 20 | 26 | 33 | 40 | 46 | 53 | 60 | 66 |

of $c$ is determined based on the previous performance study done on the SIP proxy server that includes CPU, processing speed and memory characteristics. All the incoming SIP packets are processed by different threads (allocated randomly) in parallel and forwarded to the queue until each SIP session is setup and torn down (as shown in Fig. 2.6). This process reduces the processing time of each SIP session and also increases the call arrival rate. This part of the research establishes the $M/M/c$ model as an efficient and better performance model with analytical and experimental *Data Set 2* results shown in Tables 6.1, 6.2 and Figs. 6.4–6.6.

The mean response time ($W$) and the mean number of calls ($L$) for the $M/M/c$ queuing based SIP proxy server model can be obtained from any standard work (Adan and Resing 2001; Stewart 2003) and are as follows: (For complete derivation, refer to Chap. 11)

$$W = \frac{1}{\mu} + \left[ \frac{(\frac{\lambda}{\mu})^c \mu}{(c-1)!(c\mu - \lambda)^2} \right] p_0 \tag{6.1}$$

$$L = \frac{\lambda}{\mu} + \left[ \frac{(\frac{\lambda}{\mu})^c \lambda \mu}{(c-1)!(c\mu - \lambda)^2} \right] p_0 \tag{6.2}$$

where $c$ represents the numbers of servers which is the same as number of threads that can be allotted and executed dynamically while running the SIP proxy server software. The servers provide independent and identically distributed exponential service at rate $\mu$ as shown in Fig. 6.1. $\lambda$ represents the arrival rate of requests in the system, and is the same in every state of the system. When the number of requests resident in the proxy server is greater than $c$, all the servers are busy and the mean system output rate is equal to $c$ times $\mu$. When the number of requests $n$ is less than $c$, then only $n$ out of $c$ servers are busy and the mean system output rate is equal to $n$ times $\mu$. Following usual birth-death derivation, the equilibrium probability $p_0$ of the system being idle can be obtained using

$$\left[ 1 + \sum_{n=1}^{c-1} \frac{(c\rho)^n}{n} + \frac{(c\rho)^c}{c!} \frac{1}{1-\rho} \right]^{-1} \tag{6.3}$$

where $\rho = \frac{\lambda}{c\mu} < 1$

## 6.3  Experiment Lab Setup

Configured each of 8–10 HP 7825H2 servers with the Intel dual core Pentium configured as various network elements such as User Agent Client (UAC), SIP proxy server loaded with CISCO SIP proxy server Software, User Agent Server (UAS), Linux based DNS server, CISCO Camelot call generator tool (CISCO Internal tool 2004) and Performance monitor (*Perfmon*) tool (XR Perfomon tools 2006) along with CISCO 3825 router, and a CISCO 3745 NAT server to perform all the experiments. All the equipments are configured within a local lab network (LAN environment) without any external network interferences that can impact the performance numbers. UAC and UAS are sending and receiving SIP packets as shown in Fig. 6.3. All the CISCO 7970 model SIP phones simulated by the CISCO Camelot tool used seven digits Directory Number (DN), since we used the SIP proxy server within the same domain. Refer to Chap. 12 for detailed lab setup diagrams.

## 6.4  Experiment Procedure

From UAC, SIP calls are made using CISCO Camelot simulator with different call rate of 50, 100, 150, 200, 250, 300, 350, 400, 450, 500, 600, 700, 800, 900, 1,000 cps (calls per second) were sent to the SIP proxy server, which is under test as shown in Fig. 6.3. Note: 500+ cps arrival rates are tested only for $M/M/c$ queuing model experiments, because the server utilization went above 90 % range after 450 cps in case of $M/M/1$ as shown in Table 4.1 and started losing packets (meaning Call Completion Rate goes below 100 %). SIP proxy server process the

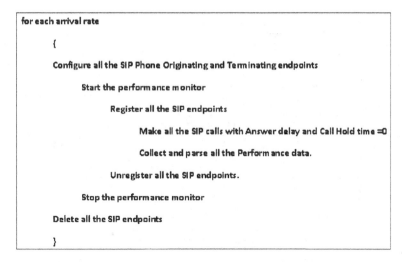

**Fig. 6.2** Lab experiment automation code flow chart

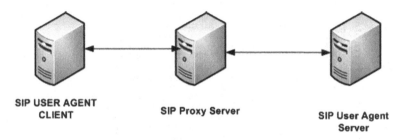

**Fig. 6.3** Experiment setup

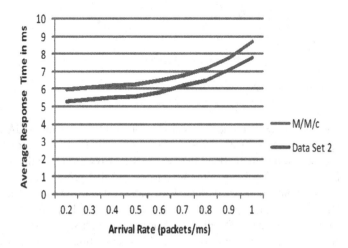

**Fig. 6.4** $M/M/c$ ART verses experimental data set 2

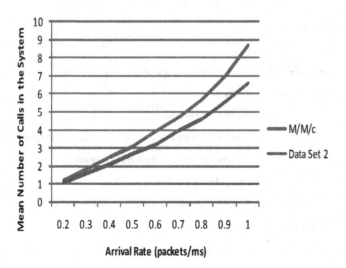

**Fig. 6.5** $M/M/c$ MNC verses experimental data set 2

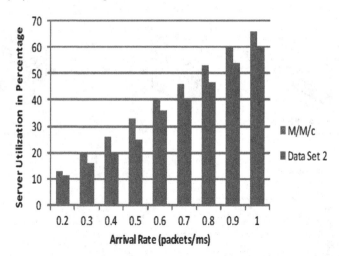

**Fig. 6.6** $M/M/c$ SU verses experimental data set 2

incoming SIP packets (SIP request message) from UAC and sent to UAS. UAS sent the responses back to UAC via SIP proxy server. Performance monitor tool (*Perfmon*) is configured with the IP address of the SIP proxy server, all the necessary counters that are already instrumented as part of the SIP proxy server software. During the experiments, data are collected for 1,000, 2,500, 5,000 and 10,000 SIP calls. Then the total mean response time data were calculated by averaging all the mean response time of each SIP call. Based on the SIP call setup ladder diagram, calculated the mean response time of all the SIP transactions between A and N. The time delays between all transactions are deducted by capturing the time stamps between actual SIP transactions. Performance monitor tool (*Perfmon*) is also CISCO home grown monitor tool that collects the traffic data. All the Camelot simulation scripts and associated library of functions are written using Tcl/Tk programming language (Fig. 6.2 *details a sample pseudo code*) and ran from CISCO home grown GUI based tool called Strauss (CISCO Internal tool 2006). Several tests are conducted for various $\lambda$ values for both the models. While making the SIP calls, "Answer ring tone delay" is set to zero, meaning when the SIP phone rings on the terminating side, the time delay to answer the phone is set to zero, assumed the phone answers immediately. Same manner, "Call Hold Time" also known as "talk time" is set to zero meaning calls will be terminated immediately after establishing the voice path for all the experiments. Also, monitored for any packet loss and validated the complete SIP call flow as shown in Fig. 2.6 for each SIP call during the experiment using "Wireshark" network protocol analyzer tool (Wireshark 2008).

All the experiments conducted similar to $M/M/1$ model with the CISCO SIP proxy servers with all the simulation tool, data collection tool, hardware, CISCO SPS software, and used automated scripts to perform the experiments.

## 6.5 Predictions of M/M/c Model

Queuing Tool Pack 4.0 queuing simulation tool (Ingolfsson and Gallop 2003) is a Microsoft Excel 2007 plug-in (QTP.exe and QTP.dll files) is used to calculate the analytical data. The plug-ins can be downloaded into Microsoft office library directory. All the queuing formula and the corresponding calculations can be accessed through the spreadsheet from the *Formulas menu* options, then select the "Queuing ToolPak" under category, and then select the appropriate queuing model. We calculated average response time, mean number of number of calls and server utilization data by providing appropriate input values such as $\lambda$, and $\mu$. and $c$ (only in case of $M/M/c$) based on the $M/M/1$ and $M/M/c$ SIP proxy server models. We considered $c = 3$ (number of threads) in our $M/M/c$ model calculations, based on the internal study done within CISCO development team on the proxy server optimal value for number of threads needed for processing the SIP packets. The predictions of $M/M/c$ model is inline or very close to the experimental *Data Set 2* in the real world experiments and in absolute numerical values.

## 6.6 Scalability of $M/M/c$ Based SPS

In this part of the research, we investigated the scalability of the SIP proxy server and extended the research to investigate additional performance characteristics of the SIP proxy server such as CPU utilization, queue size and memory utilization for the generally acceptable average Call Hold Time (CHT) and validated the results with lab experiments in the LAN environment with the single proxy server model. These new results, in addition to prior understanding of the performance characteristics of the SIP Proxy Server, further validates our $M/M/c$ model. For server utilization and queue length, there are mathematical references (Adan and Resing 2001; Stewart 2003) that are available for the $M/M/c$ based queuing model. In addition to that the memory utilization of the server is also studied only using the experimental method and presented the data in this part of research. To measure the scalability of the SPS, we measured the CPU cost, queue length and memory cost for each SIP call. For that, during the experiments, we considered the "Auto ring tone delay", meaning when the SIP phone rings on the terminating side, the time delay to answer the phone is random and the "Call Hold Time (CHT)" also known as "talk time" is set to 180 s (3 min), meaning the calls will be terminated after 3 min as generally accepted industry standards. The active SIP calls in the system can be calculated as arrival rate $(\lambda) \times CHT$. Example: For arrival rate $(\lambda)$ 100 cps, $100 \times 180 s = 18,000$ calls active in the system. The experimental data for the scalability tests of the SPS are presented in Figs. 6.7–6.15. In these graphs, Y-axis represents the memory, CPU and queue size and the x-axis represents the polling interval at 10 s per interval, meaning, every 10 s the (*Perfmon*) tool, poll the SPS server and extract the data.

- The server utilization predicted by the $M/M/c$ queueing model are in general agreement with the experimental results, as shown in Figs. 6.7–6.10. This clearly indicates that the $M/M/c$ model can successfully predict server utilization of the SPS, under varying conditions of load.
- The CPU and memory utilization increases when the arrival rate increases as predicted and it is peaked between 2.5–16 % range for the 2 CPU 4 Processor server as shown in Figs. 6.7 and 6.15 for 1,000 cps without any call loss is a good indicator how many calls that this SPS server can handle. This further validates that the multi-threaded SPS realization is scalable.
- In case of CPU and queue size data (Figs. 6.7 and 6.11), after reaching 180 s CHT, CPU curve fluctuates but in a constant and regular intervals. This indicates the multi-threaded SPS realization is more stable.
- The queue length Fig. 6.11 increases when the call rate increases and its in a very acceptable range.
- The queue sized predicted by the $M/M/c$ queueing model are in general agreement with the experimental results, as shown in Figs. 6.11–6.14. This shows that our $M/M/c$ model can successfully predict buffer occupancy of the SPS, under varying conditions of load.
- In case of memory data, another key observation, for each arrival rate after reaching 180 s interval, the curve stays flat (Fig. 6.15) because the calls that are active in the system is equal to $\lambda \times CHT$ (180 s), so number of calls active in the system for each arrival rate is a constant. This flat curve also indicates there is no memory leaks in the system.
- Memory cost average per SIP call is measured as 2.92 KB and CPU cost average cost per SIP is measured as 0.00625 % per call, which is acceptable to most

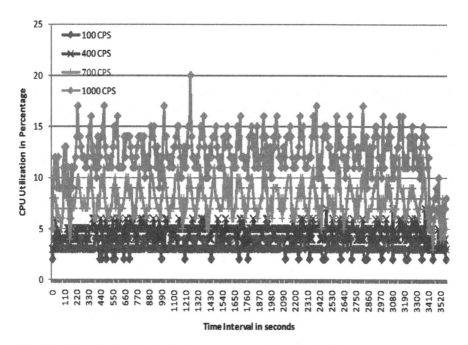

**Fig. 6.7**  CPU utilization chart (other 6 cps rates are not shown for clarity)

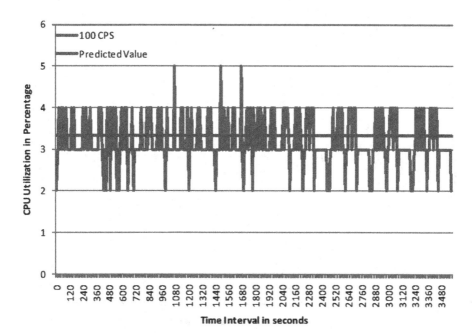

**Fig. 6.8**  CPU utilization 100 cps predicted and experimental comparison

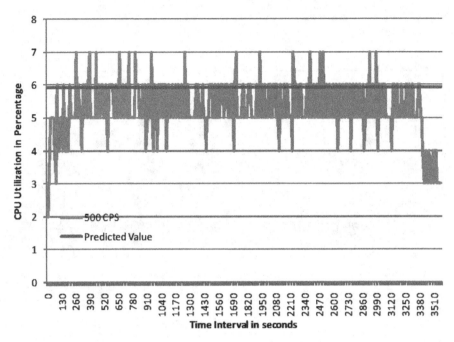

**Fig. 6.9** CPU utilization 500 cps predicted and experimental comparison

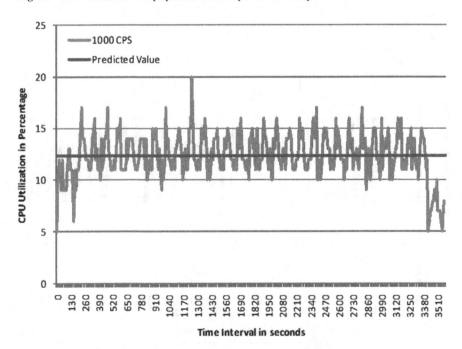

**Fig. 6.10** CPU utilization 1,000 cps predicted and experimental comparison

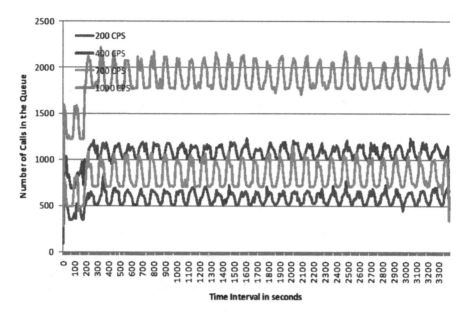

**Fig. 6.11**   Queue size chart (other 6 cps rates are not shown for clarity)

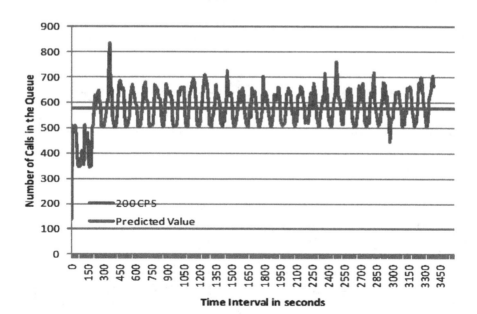

**Fig. 6.12**   Queue size 200 cps predicted and experimental comparison

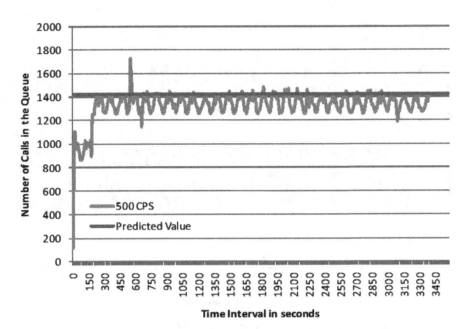

**Fig. 6.13** Queue size 500 cps predicted and experimental comparison

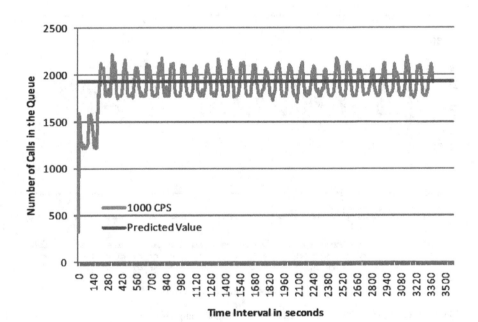

**Fig. 6.14** Queue size 1,000 cps predicted and experimental comparison

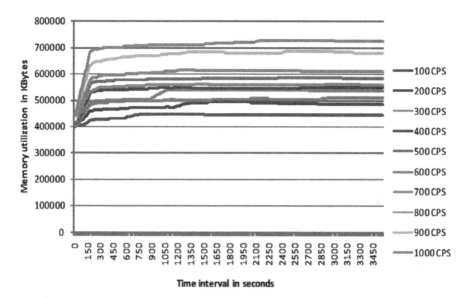

**Fig. 6.15** Memory utilization chart

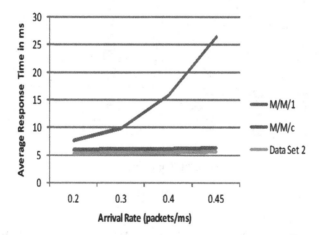

**Fig. 6.16** $M/M/1$ and $M/M/c$ ART verses experimental data set 2

industry standards. This clearly indicates the multi-threaded SPS realization is scalable and have a better performance.

• CPU cost, Memory cost and the Queue size increases linearly when the call arrival rate increases, as predicted.

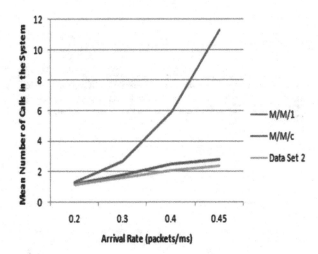

**Fig. 6.17** $M/M/1$ and $M/M/c$ MNC verses experimental data set 2

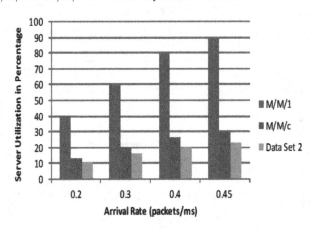

**Fig. 6.18** $M/M/1$ and $M/M/c$ SU verses experimental data set 2

## 6.7   Comparative Study on Absolute Performance of SPS

Performance of the SIP proxy servers varies from servers to servers depends on the hardware capacity, CPU, speed and other memory related parameters. The experiments are conducted with a single SIP proxy server in local LAN environment.

- In $M/M/1$ model, we noticed, there are significant differences between the analytical data versus experimental *Data Set 2* obtained from the real proxy server (see Figs. 6.7 and 6.8), likely due to several assumptions and limitations made in the analytical model. More importantly, the difference is clearly not just one of scale, i.e. tuning the single $\mu$ parameter in the $M/M/1$ model proposed in previous work would not succeed in matching the experimental data for any value of $\mu$.

- In $M/M/c$ model, the experimental results obtained is slightly better than the analytical results, approximately 10–15 % as shown in Figs. 6.7 and 6.8, since there are very limited assumptions made in our analytical model. When the call arrival rate increases, then the average response time and the mean number of calls increase. This confirms the intuition that the $M/M/c$ model more nearly represents the correct architecture of the SIP proxy server.
- In case of emulated $M/M/1$ model, there are six queuing stations processing all the incoming SIP packets, the average response time to establish each SIP session is in the range between 9.8 and 26.5 ms. Whereas in case of $M/M/c$ model with a single queue with three servers, is in the range between 6 and 6.3 ms for the same cps rates with the experiments conducted on a real SIP proxy server. This indicates $M/M/c$ model showed a significant performance improvement compared to $M/M/1$ model.
- Incase of $M/M/1$ model, the number of calls in the system (1.31–11.3 calls) is much higher than the $M/M/c$ model (1.2–2.7 calls). Again $M/M/c$ model is 75 % better than $M/M/1$ model as shown in Figs. 6.7 and 6.8.
- During the experiments, we observed that the *Data Set 2* performed much better when the arrival rate of the incoming packets increases compared to *Data Set 1*. In *Data Set 1*, maximum call rate achieved is 450 cps, whereas *Data Set 2*, the call rate achieved is 1,000 cps as shown in Figs. 6.7 and 6.8.
- The maximum call rate (in cps) that the $M/M/1$ model can process is 50 % lower than $M/M/c$ model as shown in Tables 4.1, 6.1, and 6.2.
- In the $M/M/1$ model the server utilization reaches 80 % while processing at the Call rate of 450 cps, whereas, in $M/M/c$ model, the server utilization $\rho$ stays close to 60 % while processing the Call rate of 1,000 cps as shown in the Table 4.1 and Fig. 6.9. Data presented for $\rho$ is rounded to the whole number. By comparing the average server utilization data between the two models for the same cps, $M/M/c$ model outperforms the $M/M/1$ model, meaning $M/M/c$ model can process more number of incoming SIP calls.
- In case of $M/M/c$ model, we noticed that the average response time and mean number of jobs is almost constant/slight change for the call rates up to 400 cps, since the calls are served randomly by 3 servers in the system as shown in the Figs. 6.7 and 6.8 and then increases slightly around 6 ms for higher call rates.
- Average response time to establish each SIP session obtained in the *Data Set 2* and the predicted $M/M/c$ model are between 6–8.7 and 5.3–7.8 ms respectively and that meets the ITU standards.

## 6.8 Overall Comparison of All Three Performance Models

Based on the analytical and experimental results presented in this research, $M/M/c$ queuing based SIP proxy server model outperforms the $M/D/1$ model as well as $M/M/1$ queuing based models with respect to average response time, mean number

**Table 6.3** Overall performance model comparison data

| Model | $M/M/1$ | Data set 1 | $M/D/1$ | $M/M/c$ | Data set 2 |
|---|---|---|---|---|---|
| L | 1.31–11.33 | 0.82–5.68 | 1.03–6.3 | 1.2–8.7 | 1.1–6.6 |
| W in ms | 7.63–26.47 | 5.8–13.47 | 7.16–16.16 | 6.0–8.7 | 5.3–7.8 |
| $\rho$ in percent | 40–90 | 28–75 | 40–90 | 13–66 | 11–60 |

of calls in the system and the server utilization of the system. Overall comparison results are shown in Table 6.3.

## 6.9　Concluding Remarks

Based on the measurements and analysis, this research (Subramanian and Dutta 2009b) found that the SIP Proxy Server with $M/M/c$ queuing model produces better prediction of server performance than the $M/M/1$ model with six queuing stations, proposed by authors in Gurbani et al. (2005). This work also found that the SIP Proxy Server architecture modeled by the $M/M/c$ model can scale well in terms of processing number of incoming SIP calls. This provides theoretical justification for preferring the multiple identical threaded realization of the SIP proxy server architecture as opposed to the older, thread-per-module realization. With the predicted and experimental results, we established that the average response time, mean number of calls and server utilization factor of the $M/M/c$ model can produce a more predictable with significant performance improvements and also met the ITU-T standards (as shown in Figs. 6.16–6.18). Based on the measurements and analysis, this research found that the multi-threaded SIP Proxy Server architecture can scale well in terms of processing the number of incoming SIP calls, and additional properties of such a system such as server utilization and buffer occupancy can be successfully modeled by the $M/M/c$ model. This work also establishes that the memory and CPU cost to process each SIP call is very reasonable and this clearly indicates the multi-threaded SPS can meet most of the performance and scalability standards. In the future, we intend to expand this research, considering multiple SIP Proxy Servers located in various remote locations and do a comparative study of the performance impact, when network delays are introduced into these models.

# Chapter 7
# Performance of the SPS in LAN and WAN Environment

The core idea for this part of research is to investigate the multi proxy environment, so that the SIP calls can pass through multiple proxy servers and latency is considered. We investigated the impact of Call Hold Time (CHT) in SPS performance with the help of a separate empirical study in the lab. For this study, we considered a multi-hop proxy server setup by redesigning our previous test setup with the latest IETF recommended trapezoidal SIP network architecture. The result of that investigation motivated us to further explore the SPS performance in LAN and WAN environment. The chapter provides details about our empirical study on CHT, results, empirical and predicted results on $M/M/c$ based SPS model discussed in Chap. 6.

## 7.1 Empirical Study of Call Hold Time in SPS Performance

In earlier research, all the experiments are conducted in a single proxy server setup in a LAN environment with Call Hold Time (CHT) set to 0 (meaning all the calls will be terminated immediately after connecting). But in reality, once calls are established between two SIP UA (SIP User Agents), the call will stay active for sometime before the actual termination of the call. When the calls are active in the system, the new incoming calls needs to wait in the queue for a period to get processed by the SIP proxy server and this in turn requires a longer call setup time. When we increase the CHT for a longer duration, the queue length also increases linearly and that increases the call setup time. We conducted an empirical study to validate this theory. Experiments are conducted in the lab by configuring the CHT to zero first and then increased it to different CHTs to measure the call setup time, queue length and CPU utilization of the SPSs under test. All the experiments are conducted with two proxy servers in a standard trapezoidal SIP network in a LAN environment as shown in Figs. 7.5 and 7.6.

S.V. Subramanian and R. Dutta, *Measuring SIP Proxy Server Performance*, DOI 10.1007/978-3-319-00990-2_7, © Springer International Publishing Switzerland 2013

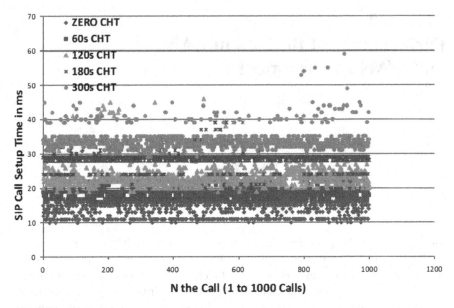

**Fig. 7.1** Call setup time for different call hold time

### 7.1.1 Key Findings from the Empirical Study

- When the CHT is set to zero, the average call set up time is 14 ms as shown in Fig. 7.1, which is very close with the practiced results obtained for the $M/M/c$ based SPS model as shown in Table 7.1.
- When the CHT is increased linearly, the call setup also increases linearly as shown in Fig. 7.1 and it shows significant difference compared to CHT set to zero.
- When the CHT is increased, the queue size also increases linearly as shown in Fig. 7.2. This intuition is, this queue size increase is causing the call setup time to increase because the incoming calls needs to wait until the active calls in the system is cleared periodically.
- Additional tests are conducted to find the optimal number for CHT for the current lab setup by increasing the CHT until we start seeing the calls are blocked, meaning the queue size (buffer) gets full, system CPU spiked up really high (close to 90 %) as shown in Fig. 7.3. The call blocking percent noticed during the experiment is from 0.1 to 1.5 % as shown in Fig. 7.4. Also noticed when the CHT is increased more number of calls are blocked or dropped before reaching the SPS.

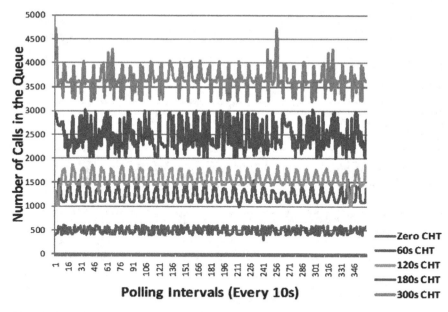

**Fig. 7.2**  Queue size for different call hold time

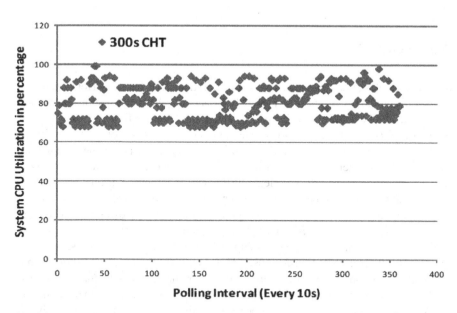

**Fig. 7.3**  System CPU snapshot when the calls are blocked

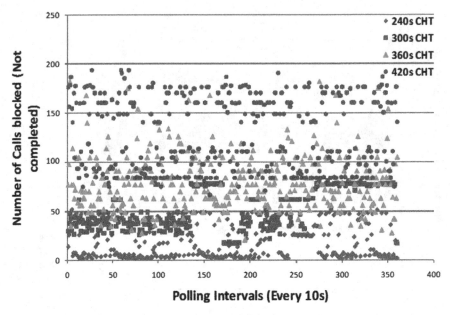

**Fig. 7.4** Call blocking chart for different call hold time

## 7.2   Motivation

Gurbani et al. (2005) came up with an analytical SIP based performance and reliability model, in which they primarily considered the mean response time and the mean number of calls in the system. They modeled a SIP proxy server as an open feed forward queuing network. The INVITE requests are forwarded upstream to the SIP proxy server from the UAC. The SIP proxy server is modeled as queuing model with six queues. In our earlier work, we proposed an analytical $M/D/1$ model, in which there are six queues in tandem to process various SIP packets with the deterministic service time (Subramanian and Dutta 2008). The SIP proxy server software module identifies and processes each SIP message based on the unique address for each session until the session is established and tore down. This creates a variation in queue occupancy at different queuing stations, while processing different SIP transactions, which increases the average response time to setup a session. In our previous work (Subramanian and Dutta 2009a), we addressed this problem, by re-designing the SIP proxy server software using multi-threaded mechanism, which can process incoming SIP packets concurrently by dynamically allocating the incoming SIP messages to various threads. We established performance measurements in our earlier research (Subramanian and Dutta 2010b) focused on a single proxy server located in the same domain. We adopted the $M/M/c$ SIP proxy server model from our previous research (Subramanian and Dutta 2009a) and enhanced it to support multiple proxy servers model. In LAN, all the network elements are configured within the local network, accessing the local DNS server for address resolution during routing, whereas in WAN, all the network elements are configured in different remote locations in different domains, accessing the public DNS server for address resolution during routing, since the calls are going through Internet. This part of research is focused on obtaining the realistic performance of SPS when the SIP calls are made in an Intranet (Example calls made within a campus) and Internet (calls made between two campuses located in different locations) setup.

The mean response time $(W)$ and the mean number of calls $(L)$ for the $M/M/c$ queuing based SIP proxy server model can be obtained from any standard work (Adan and Resing 2001; Stewart 2003) and are as follows:

$$W = \frac{1}{\mu} + \left[ \frac{(\frac{\lambda}{\mu})^c \mu}{(c-1)!(c\mu - \lambda)^2} \right] p_0 \tag{7.1}$$

$$L = \frac{\lambda}{\mu} + \left[ \frac{(\frac{\lambda}{\mu})^c \lambda \mu}{(c-1)!(c\mu - \lambda)^2} \right] p_0 \tag{7.2}$$

where $c$ represents number of servers which is same as number of threads that can be allotted and executed dynamically while running the SIP proxy server software. The servers provides independent and identically distributed exponential service at rate $\mu$ as shown in Fig. 6.1. $\lambda$ represents the arrival rate of requests in the system, and is the same in every state of the system. When the number of requests resident in the proxy server is greater than $c$, all the servers are busy and the mean system output rate is equal to $c$ times $\mu$. When the number of requests $n$ is less than $c$, then only $n$ out of $c$ servers are busy and the mean system output rate is equal to $n$ times $\mu$. Following the usual birth-death derivation, the equilibrium probability $p_0$ of the system being idle can be obtained using:

$$\left[ 1 + \sum_{n=1}^{c-1} \frac{(c\rho)^n}{n} + \frac{(c\rho)^c}{c!} \frac{1}{1-\rho} \right]^{-1} \text{where } \rho = \frac{\lambda}{c\mu} < 1 \tag{7.3}$$

If $P_1, P_2, P_3 \ldots P_n$ be the number of SIP proxy servers (hops) required to establish a complete SIP sessions, then the mean response time should be:

$$W_1 + W_2 + \ldots + W_n \tag{7.4}$$

and the mean number of calls in the system should be:

$$L_1 + L_2 + \ldots + L_n \tag{7.5}$$

## 7.3   SIP Session Setup in Different Domain

We re-designed the single server model into multi-hop IETF's standard SIP trapezoidal model with two proxy servers to establish SIP sessions in a LAN and WAN environment. SIP UA Client (UAC) and SIP UA Server (UAS) are SIP phones running a soft client that can support voice and video. Upon powering up, both users register their availability and their IP addresses with the SIP proxy server in the Internet Service Provider (ISP) network. In Fig. 7.1, UAC initiates the calls to UAS attached to SPS2 in domain B. To do so, it communicates with SPS1 in its domain (domain A). SPS1 forwards the request to the proxy for the domain of the called party (domain B), which is SPS2. SPS2 forwards the call to the called party, UAS. As part of this call flow, SPS1 needs to determine a SPS for domain B. In case of WAN, SPS1 access both SRV (Gulbrandsen et al. 2000) and NAPTR (Mealling and Daniel 2000) records to get the routing information on a remote proxy server and local DNS server to resolve the routing in LAN environment. SPS1 specifically needs to determine the IP address, port, and transport protocol for the server in domain B. The Domain B SPS delivers UACs invitation to UAS, who forwards his/her acceptance along the same path the invitation travelled. To terminate the call either UAC or UAS will send a BYE message. Figure 7.1 details the exchange of SIP requests and response messages between UAC and UAS through SIP proxy servers.

## 7.4   Experiment Lab Setup

We used standard SIP trapezoidal network setup as our testbed as shown in Fig. 7.5. We configured each of 8-10 HP 7825H2 servers with the Intel dual core Pentium configured as various network elements such as User Agent Client (UAC), SIP proxy server loaded with CISCO SIP proxy server Software, User Agent Server (UAS), Linux based DNS server, CISCO Camelot call generator (2004) and Performance Monitor (*Perfmon* tools (2006)) tool along with CISCO 3825 Router, and a CISCO 3745 NAT server to perform all the experiments. All the CISCO 7970 model SIP phones simulated by the CISCO Camelot tool used seven digit Directory Number (DN), since we used the SIP proxy server for setting up the SIP calls in a LAN and E.164 11 digit Directory Number (DN) for the calls going through WAN (Internet). For the LAN calls are set up within the local CISCO campus and for the WAN calls are setup between Raleigh, North Carolina to Dallas, Texas and Raleigh to SanJose, California. LAN calls used the local DNS server to route the calls, whereas WAN used the public DNS server to route the calls. To emulate WAN environment inside a test lab (LAN), considered the following: (1) Throttle bandwidth by recreating a WAN link by restricting the throughput on the LAN link; (2) Network Latency; and (3) Network errors such as packet loss and packet corruption. Used a open source tool, WAN emulator (*WANem*) tool (2008) to test the SIP calls going through WAN. Refer to Fig. 12.4 for complete lab setup for the LAN and WAN experiments.

## 7.5   Experiment Procedure

From UAC, SIP calls are made using CISCO Camelot simulator with different call rates of 25, 50, 100, 150, 200 cps (calls per second) sent to the SIP proxy servers, which is under test as shown in Fig. 7.5. SIP proxy server process the incoming SIP packets (SIP request message) from UAC and sent to the proxy server attached to UAS. UAS sent the responses back to UAC through two SIP proxy servers. Performance monitor tool (*Perfmon* tools (2006)) is configured with the IP address of the SIP proxy Severs, all the necessary counters that are already instrumented as part of the SIP proxy server software. During the experiments, data were collected for 1,000, 2,500, 5,000 and 10,000 SIP calls. Then the total mean response time data were calculated by averaging all the mean response time of each SIP Call. All the Camelot simulation scripts and associated library of functions are written using Tcl/Tk programming language and ran from CISCO home grown GUI based tool called Strauss tool (2006). Several tests are conducted for various $\lambda$ values for both the models. While making the SIP calls, "Answer ring tone delay" is set to random, meaning when the SIP phone rings on the terminating side, the time delay to answer the phone is set to 1–5 rings, assumed the phone answers randomly anywhere between 1 and 5 rings. In the same manner, "Call Hold Time" (CHT) also known as "talk time" is set to random between 30 and 180 s meaning calls will be terminated after 30–180 s after establishing the voice path for all the experiments. Also, we monitored for any packet loss and validated the complete SIP call flow as shown in Fig. 7.6 for each SIP call during the experiment using "Wireshark"

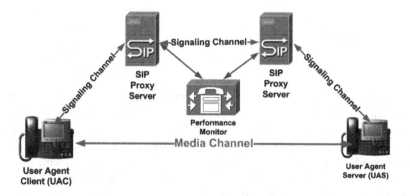

**Fig. 7.5** Experiment setup

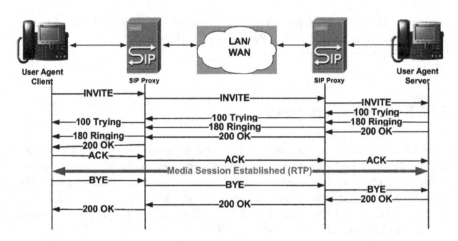

**Fig. 7.6** SIP actual message transaction diagram

**Fig. 7.7** WAN analyzer
sample output

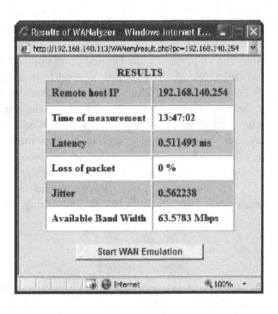

network protocol analyzer tool (Wireshark 2008). Used the *WANem* tool (2008)
to emulate the WAN environment. Start the *WAN analyzer* option first from the
*WANem* GUI to measure available bandwidth, latency, loss and jitter of a wide
area network in a few seconds, given the IP Address of a remote host as input.
Sample WAN analyzer output is shown in Fig. 7.7. Measuring the characteristics
help in giving realistic input to the WANem. WANem is a commonly used tool to
emulate WAN environment in the developer community to conduct the performance
tests and produce accurate results. Sample packet flow with WAN setup is shown in
Fig. 7.8.

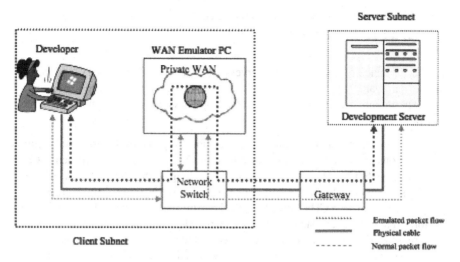

**Fig. 7.8** WANem WAN packet flow sample

**Table 7.1** Two proxy server $M/M/c$ performance model predicted results

| $\lambda$ (calls/s) | 25 | 50 | 100 | 150 | 200 |
|---|---|---|---|---|---|
| L | 0.6 | 1.2 | 1.5 | 2 | 2.4 |
| W in ms | 3 | 5 | 8 | 10 | 12 |
| $\rho$ in percent | 10 | 16 | 19 | 22 | 26 |

## 7.6 Predicted Results

Queuing Tool Pack 4.0 queuing simulation tool (Ingolfsson and Gallop 2003) is a Microsoft Excel 2007 plug-in (QTP.exe and QTP.dll files), which can be downloaded into Microsoft office library directory. All the queuing formula and the corresponding calculations can be accessed through the spreadsheet from the *Formulas menu* options, then select the "Queuing ToolPak" under category, and then select the appropriate queuing model. We calculated average response time, mean number of number of calls and server utilization data by providing appropriate input values such as $\lambda$, $\mu$ and $c$ for the $M/M/c$ SIP proxy server models. We considered $c = 3$ (number of threads) in our $M/M/c$ model calculations, based on the internal study done within CISCO development team on the proxy server optimal value for number of threads needed for processing the SIP packets. Predicted results are shown in Table 7.1. The predicted result shown is the sum of two-proxy servers result in the SIP network. Server utilization is the average for both the SIP proxy servers.

## 7.7   Comparative Study on Absolute Performance of a SIP Proxy Server

The performance of the SIP proxy servers varies from servers to servers depends on the hardware capacity, CPU, speed and other memory related parameters. The experiments are conducted with two SIP proxy servers in LAN and WAN environments.

- Experiment result indicates more realistic data compared to the predicted results for various call arrival rates as shown in Table 7.1. During the experiments, Call Hold Time (CHT) played a significant role along with other network element delays in processing the call in both LAN and WAN setup. This justifies our empirical studies on CHT impacts on SPS performance (addressed in Sect. 7.1.1). Noticed that the queue size increased linearly when the Call Hold Time (CHT) is increased. Also, answer ring tone delay contributed approximately 2.5 ms for the experimental results along with other network delays in processing the call.
- SIP call setup time in WAN (76–125 ms) is significantly higher than the call setup time in LAN (18–34 ms) for various call arrival rates, due to expensive lookup and routing, since the calls are going through public DNS server in WAN compared to local DNS incase of LAN as shown in Figs. 7.9 and 7.10. Since the CHT is set at random intervals between 30 and 180 s for each call, the number of calls active in the system should be $\lambda \times CHT$. The queue size increased linearly

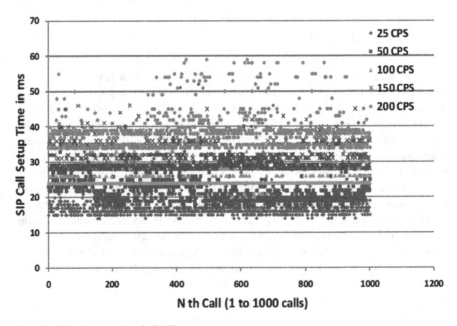

**Fig. 7.9**  SIP call setup time in LAN

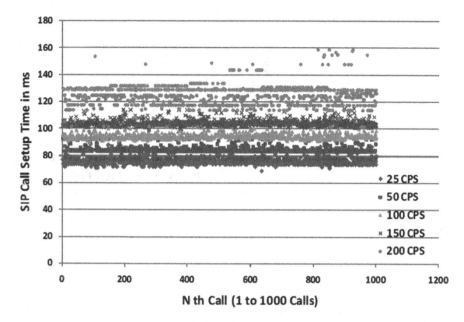

**Fig. 7.10**  SIP call setup time in WAN

**Fig. 7.11**  Originating SPS QSize in LAN

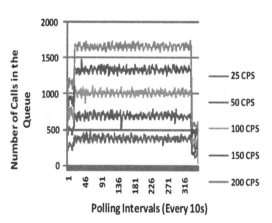

based on the call arrival rate and that increases the call setup time. Another important reason for the higher call setup time in WAN, is the lookup time to obtain the routing information is longer compared to LAN, where the lookup is done in a local DNS server.

- Observed that the performance measurements data such as CPU, Memory and Queue sizes are 10–15 % higher at the originating proxy server compared to terminating proxy server and all the data indicates linearity when the call arrival rates are increased from 25 to 200 cps as shown in Figs. 7.11–7.22.

**Fig. 7.12** Terminating SPS
QSize in LAN

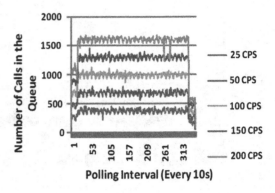

**Fig. 7.13** Originating SPS
QSize in WAN

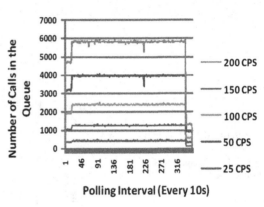

**Fig. 7.14** Terminating SPS
QSize in WAN

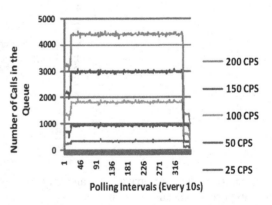

- Queue size is increased linearly for various call rates in both LAN and WAN but the actual queue size is significantly higher in WAN compared to LAN as shown in Figs. 7.11–7.14. In case of LAN, the queue size is in the range of 250–1,800 calls for various call rates whereas 500–5,800 calls in case of WAN.
- Server CPU utilization is 10–15 % higher at the originating proxy server than the termination proxy server in both LAN and WAN as shown in Figs. 7.19–7.22 due

**Fig. 7.15** Originating SPS
memory utilization in LAN

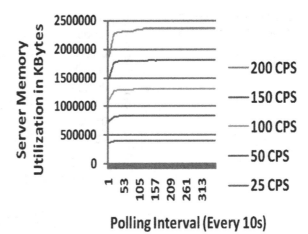

**Fig. 7.16** Terminating SPS
memory utilization in LAN

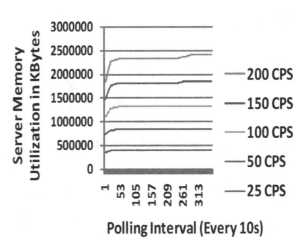

**Fig. 7.17** Originating SPS
memory utilization in WAN

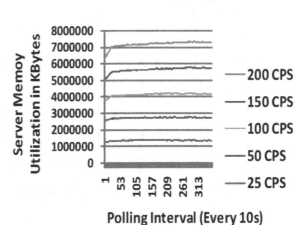

**Fig. 7.18** Terminating SPS
memory utilization in WAN

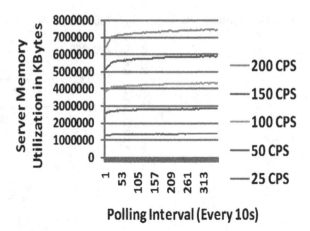

**Fig. 7.19** Originating SPS
CPU utilization in LAN

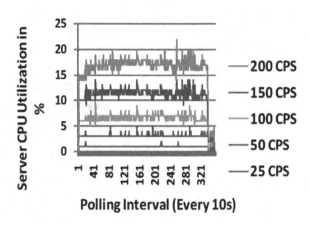

**Fig. 7.20** Terminating SPS
CPU utilization in LAN

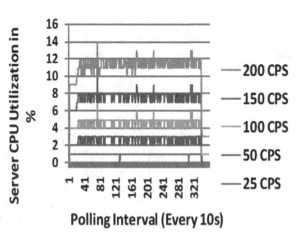

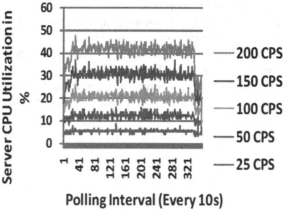

**Fig. 7.21** Originating SPS CPU utilization in WAN

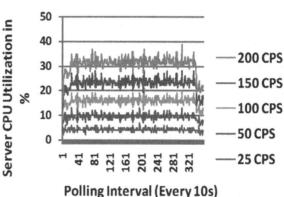

**Fig. 7.22** Terminating SPS CPU utilization in WAN

to more processing at the originating proxy server. CPU utilization of the proxy server in WAN (up to 40 %) almost doubled compared to LAN (up to 20 %).

- Server memory footprint indicates server memory utilization is significantly higher when the calls were made through WAN compared to calls went through LAN as shown in Figs. 7.15–7.18. The flat memory curve for various call rates after the number of active calls reached the limit ($\lambda \times CHT$) indicates that there are no possible memory leaks in the SIP proxy server software.

## 7.8  Concluding Remarks

This part of research establishes that there are significant performance impacts on the SIP proxy servers when the calls were through WAN verses calls made through LAN environments. Though it is obvious in theory performance of the SPS will be higher when the calls are made through public Internet, to establish that for fact we conducted several real time experiments in the lab during this work, collected the data and established the benchmark data for SIP call setup time, queue size, memory and CPU utilization (Subramanian and Dutta 2010b). Since, we used the same $M/M/c$ based SPS model proposed in our earlier work (Subramanian and Dutta 2009a). This provides sufficient justification for preferring the multiple identical threaded realization of the SIP proxy server architecture to process SIP calls in the LAN and WAN setup, thread-per-module realization. Intend to extend the studies on performance impacts of implementing security in SIP.

# Chapter 8
# SPS Performance Overheads with SIP Security

Security to SIP network is of foremost important, since SIP is deployed in a large scale in various industries, universities, Department of Defense, banks, Wall Street and other areas where they use Internet telephony. Very little research has been focused so far in evaluating the SPS when different non-secure and secure transport protocols are used to transport the packet, while establishing a SIP session. This chapter is mainly focused on SPS performance overhead to the service providers when SIP calls are made using secured and non-secured transport protocols in the LAN environment.

## 8.1 Motivation

Session Initiation Protocol has become the call control protocol of choice for VoIP networks because of its open and extensible nature. However, the integrity of call signaling between sites is of utmost importance, and SIP is vulnerable to attackers when left unprotected. In Session Initiation Protocol (SIP) based network, there are two kinds of possible threats – external and internal. External threats are attacks launched by someone not participating in the message flow during a SIP-based call (Dierks and Allen 1999). External threats usually occur when the voice and signaling packets traverse untrustworthy boundaries, and may involve third-party networks when the call traffic is transferred from device to device, or from participant to participant. Internal threats are much more complex because they are usually launched by a SIP call participant. Because a SIP call participant launches internal attacks, the trust relationship is defied. Usually, endpoints within the enterprise are administratively controlled and secured behind firewall protection so they are not expected to be capable of launching an attack. Once the trust relationship is defied and an endpoint acts in a hostile manner, identifying the source of the attack and determining the scope of the remedy becomes very difficult (Sulaiman et al. 2008).

S.V. Subramanian and R. Dutta, *Measuring SIP Proxy Server Performance*,
DOI 10.1007/978-3-319-00990-2_8,
© Springer International Publishing Switzerland 2013

Some of the possible threats to the SIP network are: (1) Denial-of-service (DoS) attacks: Prevention of access to a network service by bombarding SIP proxy servers or voice-gateway devices on the Internet with inauthentic packets; (2) Eavesdropping: Unauthorized interception of voice packets or Real-Time Transport Protocol (RTP) media stream and decoding of signaling messages; (3) Packet spoofing: Impersonation of a legitimate user sending data; (4) Replay: The retransmission of a genuine message so that the device receiving the message reprocesses it; and (5) Message integrity: Ensuring that the message received is the same as the message that was sent (Arkko et al. 1999). SIP proxy server software is implemented with encrypted mechanisms like SRTP to send the encrypted data using CSeq and Call-ID headers (Dierks and Allen 1999). SIP proxy server also implemented with participant address authentication mechanisms to address the threats.

The TLS protocol allows applications to communicate across a network in a way that is designed to prevent eavesdropping, tampering, and message forgery. TLS uses cryptography to provide endpoint authentication and communications privacy over a network. For secure SIP communications, Rosenberg et al. (2002) defines the SIPS Uniform Resource Identifier (URI), used as HTTPS is used for secure HTTP connections. The SIPS URI ensures that SIP over TLS is used between each pair of hops to validate and secure the connection, and provide a secure endpoint-to-endpoint connection. In a secure SIP session, the SIP user agent client contacts the SIP proxy server requesting a TLS session (Dierks and Allen 1999). The SIP proxy server responds with a public certificate and the SIP user agent then validates the certificate. Next, the SIP user agent and the SIP proxy server exchange session keys to encrypt or decrypt data for a given session. From this point, the SIP proxy server contacts the next hop and similarly negotiates a TLS session, ensuring that SIP over TLS is used end-to-end.

The main contribution to this part of research is to evaluate the performance impacts on the SIP proxy server when SIP calls were made using Non-Secure UDP, Non-Secure TCP, Secure-TLS-authentication (meaning the secured key exchanges between SIP entities via SIP proxy server), and Secure TLS-Encrypted (where the entire packet is encrypted with the appropriate key information) as transport protocols. During handshake SPS uses cryptographic key exchanges between SIP entities. SPS software supports

```
AES-CM-128-HMAC-SHA1-80, AES-CM-128-HMAC-SHA1-32
```

crypto suites. The SIP proxy server needs to do additional processing for each of this transport protocol options. For simplicity, we used TLS-auth and TLS-encr for TLS-authentication and TLS-encryption throughout this research. For both secured and non-secured modes, we adopted the $M/M/c$ based queuing model SIP proxy servers that we proposed in Subramanian and Dutta (2009a). Sample SIP secure handshake between an offerer and answerer is shown below.

Offerer sends:

```
 INVITE sip:bob@ua2.example.com SIP/2.0
To: <sip:bob@example.com>
From: "Alice"<sip:alice@example.com>;tag=843c7b0b
Via: SIP/2.0/TLS proxy.example.com;
branch=z9hG4bK-0e53sadfkasldk
Via: SIP/2.0/TLS ua1.example.com;
branch=z9hG4bK-0e53sadfkasldkfj
Record-Route: <sip:proxy.example.com;lr>
Contact: <sip:alice@ua1.example.com>
Call-ID: 6076913b1c39c212@REVMTEpG
CSeq: 1 INVITE
Allow: INVITE, ACK, CANCEL, OPTIONS, BYE, UPDATE
Max-Forwards: 69
Identity: CyI4+nAkHrH3ntmaxgr01TMxTmtjP7MASwli
NRdupRI1vpkXRvZXx1ja9k3W+v1PDsy32MaqZi0M5
WfEkXxbgTnPYW0jIoK8HMyY1VT7egt0kk4XrKFC
HYWGCl0nB2sNsM9CG4hq+YJZTMaSROoMUBhikVIj
nQ8ykeD6UXNOyfI=
Identity-Info: https://example.com/cert
Content-Type: application/sdp
Content-Length: xxxx
Supported: from-change

v=0
o=- 1181923068 1181923196 IN IP4 ua1.example.com
s=example1
c=IN IP4 ua1.example.com
a=setup:actpass
a=fingerprint: SHA-1 \
  4A:AD:B9:B1:3F:82:18:3B:54:02:12:DF:3E:5D:49:
  6B:19:E5:7C:AB
t=0 0
m=audio 6056 RTP/AVP 0
a=sendrecv
a=tcap:1 UDP/TLS/RTP/SAVP RTP/AVP
a=pcfg:1 t=1
```

Answerer replies:

```
SIP/2.0 200 OK
To: <sip:bob@example.com>;tag=6418913922105372816
From: "Alice" <sip:alice@example.com>;tag=843c7b0b
```

```
Via: SIP/2.0/TLS proxy.example.com:5061;
branch=z9hG4bK-0e53sadfkasldk
Via: SIP/2.0/TLS ua1.example.com;
branch=z9hG4bK-0e53sadfkasldkfj
Record-Route: <sip:proxy.example.com;lr>
Call-ID: 6076913b1c39c212@REVMTEpG
CSeq: 1 INVITE
Contact: <sip:bob@ua2.example.com>
Content-Type: application/sdp
Content-Length: xxxx
Supported: from-change

v=0
o=- 6418913922105372816 2105372818 IN IP4
ua2.example.com
s=example2
c=IN IP4 ua2.example.com
a=setup:active
a=fingerprint: SHA-1 \
  FF:FF:FF:B1:3F:82:18:3B:54:02:12:DF:3E:
  5D:49:6B:19:E5:7C:AB
t=0 0
m=audio 12000 UDP/TLS/RTP/SAVP 0
a=acfg:1 t=1
```

## 8.2   UDP/TCP Based Non-secure SIP Session Setup

SIP UAs register with a proxy server or a registrar. Proxy servers then act as an intermediary for SIP calls. The UAC sends an INVITE to its proxy server. In this INVITE, the Request-URI field contains the address of the called phone number as part of the SIP address. SDP information is included with this INVITE. The proxy server creates a new INVITE, copying the information from the old INVITE, but replacing the Request-URI with the address of the remote proxy attached to the UAS. When UAS receives the INVITE, it initiates a call setup with the UAC. It sends a SIP response 100 (Trying) to the proxy server attached to UAS, which then forwards that to the proxy server attached to UAC and then to UAC. When UAS receives the alerting message, it sends a SIP 180 (Ringing) message to the proxy server. The proxy server sends the same message to the UAC. When the end user picks up the phone, UAS then sends a SIP 200 (OK) response to the proxy server, which sends it to the UAC. SDP information for the remote end is included in this OK response. The SIP UAC acknowledges the OK response with an ACK message, and a two-way RTP stream is established between the UAC, proxy servers, and the UAS. When the UAC hangs up, it exchanges SIP BYE and OK signals with UAS

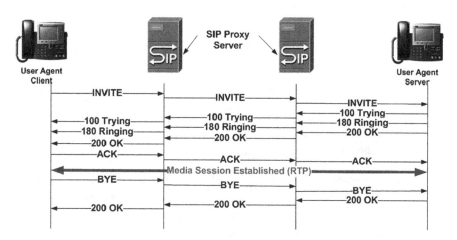

**Fig. 8.1** Non-secure SIP message transactions diagram

and UAS terminates the call with the UAC (Rosenberg et al. 2002). Figure 8.1 details the exchange of SIP requests and response messages between UAC and UAS through the intermediate SIP proxy servers.

## 8.3  TLS Based Secured SIP Session Setup

To establish a TLS handshake between the UAC and UAS, there are several additional implementations such as authentication and encryption algorithms that are required at the SIP proxy server software to process additional SIP message exchanges between UAC and UAS. The TLS handshake (Kaji et al. 2006) is implemented in three phases: (1) Authentication phase – during authentication phase SIP proxy server authenticates UAC and UAS. If authentication is success, SIP proxy server establishes and keeps SIP sessions between UAC and UAS. Both UAC and UAS register security policies with the SIP proxy server via the SIP session; (2) Key generation and distribution phase – during the key distribution phase SIP proxy server generates a security key for TLS session between UAC and UAS and distributes it to UAC and UAS through the SIP sessions. This phase is triggered when UAC tries to communicate with UAS; (3) Security Authority (SA) confirmation phase (Dierks and Allen 1999) – during SA confirmation phase, UAS and UAC confirm if they share a security association. If all the phases are executed successfully, UAC and UAS can establish a TLS session and start communicating securely. Figure 8.2 details the exchange of SIP requests and response messages between UAC and UAS through the intermediate SIP proxy servers. SRTP is a profile of RTP that is designed to provide security for RTP and it can be used for encryption, message authentication/integrity and replay protection of RTP at the media channel (Alexander et al. 2009).

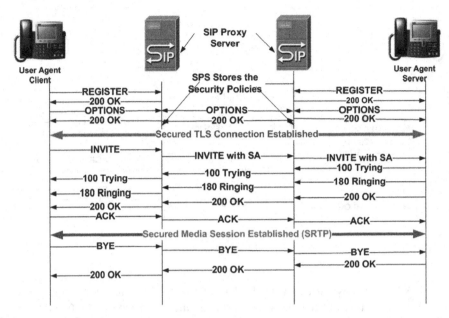

**Fig. 8.2** Secured SIP message transactions diagram

## 8.4   Experiment Setup

We used the IETF standard SIP trapezoidal network setup as our testbed as shown
in Fig. 8.3. Configured 8-10 HP 7825H2 servers with the Intel dual core Pentium
configured as various network elements such as User Agent Client (UAC), User
Agent server (UAS), Linux based Domain Name System (DNS) server, CISCO
Camelot call generator tool tool (2004) and Performance Monitor (*Perfmon* tools
(2006)) tool along with CISCO 3825 Router, and a CISCO 3745 Network Address
Translation (NAT) server to perform all the experiments. We configured 2 HP 7845
H2 servers as SIP proxy servers loaded with CISCO SIP proxy server software.
All the equipments are configured within a local lab network (LAN environment)
without any external network interferences that can impact the performance numbers. UAC and UAS are sending and receiving SIP packets as shown in Figs. 8.1
and 8.2. We used the latest SIP proxy server software implemented with multi-
threaded mechanism and followed the current software architecture. All the CISCO
7970 model SIP phones are simulated by the CISCO Camelot tool used seven digit
Directory Number (DN), since we used one SIP proxy server attached to UAC and
one SIP proxy server attached to UAS. Refer to Fig. 12.4 for complete lab setup for
the LAN experiments.

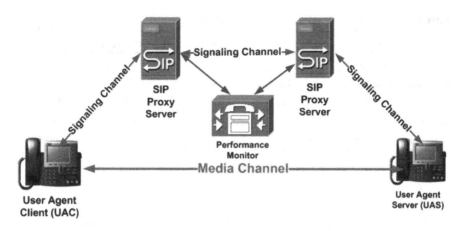

**Fig. 8.3** Experiment setup

## 8.5 Experiment Procedure

From UAC, SIP calls are made using CISCO Camelot simulator with different call rates (calls per second) sent to the SIP proxy server attached to UAC, which is under test as shown in Fig. 8.3. SIP proxy server processes the incoming SIP packets (SIP request message) from UAC and then forwards the requests the SIP proxy server attached to UAS, and SIP proxy server forwards it to UAS. UAS sends the responses back to the proxy server attached to UAS, then the responses are forwarded back to UAC through the proxy server attached to UAC. Performance monitor tool (*Perfmon* tools (2006)) is configured with the IP address of the SIP proxy servers where all the necessary counters are instrumented as part of the SIP proxy server software. During the experiments, data are collected for 1,000, 2,500, 5,000 and 10,000 SIP calls. Then the total mean response time data were calculated by averaging all the mean response time of each SIP call. Mean response time of all the SIP transactions are calculated based on the SIP call setup ladder diagrams as shown in Figs. 8.1 and 8.2. Performance monitor tool (*Perfmon* tools (2006)) is also a CISCO home grown monitor tool that collects the traffic data. All the Camelot simulation scripts and associated library of functions are written using Tcl/Tk programming language and ran from CISCO internal Graphical User Interface (GUI) based tool called Strauss tool (2006). Also, we monitored for any packet loss and validated the complete SIP call flow as shown in Figs. 8.1 and 8.2 for each SIP call during the experiment using "Wireshark" network protocol analyzer tool (Wireshark 2008). New automation Tcl/Tk scripts are developed to test the different transport protocols meaning additional parameters are added to the existing scripts to perform TLS-auth and TLS-encr modes. To measure the scalability of the SPS, we measured the CPU cost, queue size and memory cost for each SIP call. During the experiments, we considered the "Auto ring tone delay" as set to random anywhere between 1 and 5 rings, meaning when the SIP phone rings on the terminating side, the time delay to

**Table 8.1** SIP proxy server performance comparison chart

| Transport protocols | UDP | TCP | TLS-auth | TLS-encr |
|---|---|---|---|---|
| Mean call setup time (ms) | 19 | 26 | 34 | 38 |
| Mean memory (KBytes) | 440,283 | 515,566 | 631,353 | 695,086 |
| Memory cost/call (KBytes) | 147 | 172 | 211 | 232 |
| Mean CPU (%) | 30 | 35 | 41 | 45 |
| Mean queue size (calls) | 1,202 | 1,910 | 2,495 | 2,886 |

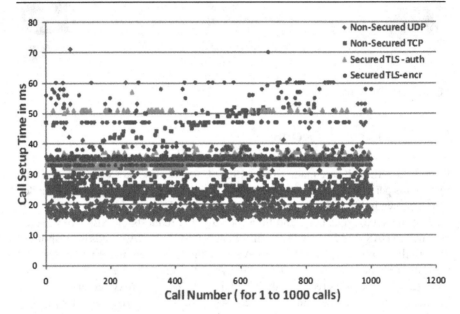

**Fig. 8.4** Non-secure and secure SIP calls setup time

answer the phone is random and the "Call Hold Time (CHT)" also known as "talk time" is set to 180 s (3 min), meaning the calls will be terminated after 3 min as generally accepted industry standards. The active SIP calls in the during the system can be calculated as arrival rate $(\lambda) \times CHT$. Example: For arrival rate $(\lambda)$ 100 calls per second, $100 \times 180 \text{ s} = 18{,}000$ calls active in the system. Results obtained from all the experiments are presented here as *Data Set 4* in Table 8.1 and Figs. 8.4–8.7.

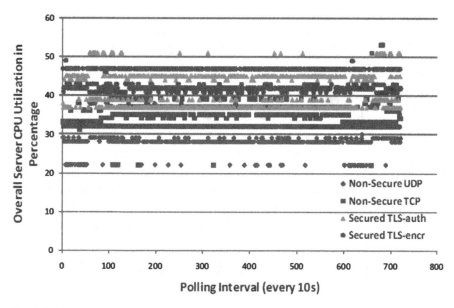

**Fig. 8.5** Non-secure and secure SPS server CPU utization

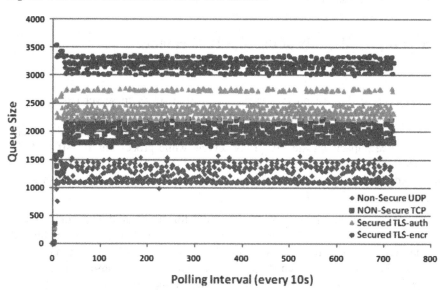

**Fig. 8.6** Non-secure and secure SPS server queue size

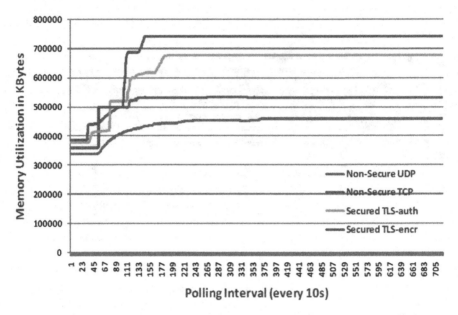

**Fig. 8.7** Non-secure and secure SPS server memory utilization

## 8.6 Empirical Study of the SPS Performance with Non-secure and Secure Transport Protocols

Performance of the SIP proxy servers varies from server to server based on the hardware capacity, CPU, speed and other memory related parameters. Performance results comparisons are shown in Table 8.1. Results clearly indicate significant performance impacts on the SIP proxy servers while processing the SIP calls in secured modes compared to non-secured modes due to additional SIP messages that needs to be processed in secured mode to accommodate the TLS handshake to happen.

- SIP call setup time is increased by 30–40% for the secured TLS compared to non-secured TCP and UDP as shown in Fig. 8.4, mainly because of additional message processing by the SPS as shown in Figs. 8.1 and 8.2. There is an increase of 4 % between TLS-authentication and TLS-encryption because additional processing of encrypting of packets before transmitting. Also SPS is implemented with a separate authentication and encryption algorithms.
- Server memory utilization is up by 35 % when SIP uses TLS when compared to non-secured TCP or UDP as shown in Fig. 8.5. The curve flattens out and almost become a straight line after the number of active calls reached the required limit $(\lambda \times CHT)$, indicates there is no memory leaks in the SPS software.

- In case of secured SIP transactions in establishing the calls, server CPU utilization is increased between 28 and 35 % compared to non-secured SIP transactions using TCP or UDP as shown in Fig. 8.5.
- The queue size increased linearly for all the transport protocols from UDP to TLS-encr as shown in Table 8.1 mainly because the CHT is 180 s, the number of calls active in the system should be $\lambda \times CHT$ as shown in Fig. 8.6.

## 8.7 Concluding Remarks

Based on the measurements and analysis, this part of our research establishes that there are significant SIP proxy server performance overheads to process additional SIP transactions, establishing secured TLS connections and setup SRTP sessions compared to RTP sessions using SIP. Even though, it is obvious that providing security in SIP based network will impact performance, this research work establishes a benchmark from the data obtained from experiments conducted in a lab setup with high-end proxy servers and tools (Subramanian and Dutta 2010a). SIP secured transactions are widely used in banks, stock markets, military purposes and other secured environments. It is possible to expand our research to study the performance impacts on the SIP proxy server in WAN environment with a similar experimental setup.

# Chapter 9
# Statistical Analysis of Experimental Data Sets

In applications, it is often important to forecast the value of a random variable Y (which can be difficult or impossible to measure directly) using the observations of another random variable X (which can be more accessible for direct measurements). An important result of the probability theory is that when the two random variables are normal the forecast of Y in terms of X that minimizes the mean-square error is linear:

$$Y' = a + bX \tag{9.1}$$

Often, even for non-normal random variables, a practically useful forecast can be obtained using a linear function of the predictor X. This explains the importance of linear regression analysis, one of the most common tools of applied statistics. In this chapter, presented the statistical analysis of all the experimental data sets to identify any significant differences (statistically). Regression analysis is a statistical technique used to quantify the apparent relationships between one or more independent variables and a dependent variable. That is, regression analysis provides a model for understanding how isolated changes in the independent variables affect the dependent variable's value. The goal for regression analysis is to find the equation that produces the best such line, with "best" defined as being that line such that the sum of the squares of the residuals – the difference between the actual value of the dependent value and the value estimated by the regression model is minimized. Regression analysis allows us to define the model in terms of the simple slope-intercept form of a linear equation:

$$y = mx + b \tag{9.2}$$

where "m" is the slope of the line (the change in Y per unit change in X), and "b" is the "intercept" term, or the value for Y when $X = 0$. Similarly regression equation can be written as:

$$y = r_0 + R_1 X \tag{9.3}$$

S.V. Subramanian and R. Dutta, *Measuring SIP Proxy Server Performance*,
DOI 10.1007/978-3-319-00990-2_9,
© Springer International Publishing Switzerland 2013

By convention, in regression analysis, the slope of the line is often referred to as the "regression coefficient", represented by the upper case R with subscript of 1, and the intercept term is often represented by the lower case r with s subscript of 0.

Point estimation from regression analysis is given by:

$$\widehat{Y}_i = r_0 + R_1 X_i + re_i \qquad (9.4)$$

In that equation, the regression coefficient (or slope, m) and the intercept (b) are givens; substituting the given value for X then determines the estimated value for Y. The r represents the residual, or error term: the estimated value may differ from an actual, observed value for Y at that given value of X.

Analysis of variance, or ANOVA, is a powerful statistical technique that involves partitioning the observed variance into different components to conduct various significance tests. Total variance of the obtained experimental data can be calculated using the formula. Additional references such as formulae and definitions are available in Chap. 14. Used Microsoft Excel spreadsheet and installed the statistical data analysis "toolpak" plug-in to perform the regression tests. By providing appropriate input values to this tool, the following data are obtained.

## 9.1   Interpretation of Linear Regression Analysis Results

This section details how to interpret the statistical results for all the experimental data sets shown from Sect. 9.2.

- $R^2 = (Multiple R)^2 = R^2 = 1 -$ Residual SS/Total SS = Regression SS/Total SS
- R = correlation coefficient
- Adjusted $R^2 = 1 - $ (Total df/Residual df) (Residual SS/Total SS)
- $Standard\,Error = (Residual\,MS)^{0.5}$
- ANOVA = ANalysis Of VAriance
- Regression df = regression degrees of freedom = number of independent variables (factors) in Eq. 9.4.
- Regression SS = Total SS – Residual SS
- Regression MS = Regression SS/Regression df
- Regression F = Regression MS/Residual MS
- Significance F = FDIST(Regression F, Regression df, Residual df) = Probability that Eq. (9.4) does NOT explain the variation in y. This is based on the F probability distribution. If it's not less than 0.1 (10%) you do not have a meaningful correlation.
- Residual df = residual degrees of freedom = Total df – Regression $df = n - 1 - $ *number of independent variables* $(X_i)$
- Residual SS = sum of squares of the differences between the values of $Y_i$ predicted by Eq. 9.4 and the actual values of Y. If the data exactly fit Eq. 9.4, then Residual SS would be 0 and $R^2$ would be 1.

- Residual MS = mean square error = Residual SS/Residual df
- *Total df = total degrees of freedom = n − 1*
- Total SS = the sum of the squares of the differences between values of Y and the average Y
  $$= (n − 1) * (standard\,deviation\,of\,Y)^2$$
- Coefficients = values of $re_i$ which minimize the Residual SS (maximize R2). The Intercept Coefficient is $r_0$ in Eq. 9.4.
- Standard error = $(Residual\,MS\,using\,only\,the\,Coefficient\,for\,that\,row)^{0.5}$
- t Stat = Coefficient for that variable/Standard error for that variable
- P-value = TDIST(—t Stat—, Residual df, 2) = the Student's t distribution two-tailed probability
- If one divides this by 2, it is the probability that the true value of the coefficient has the opposite sign to that found. You want probability to be small in order to be sure that this variable really influences y, certainly less than 0.1 (10 %).
- There is a 95 % probability that the true value of the coefficient lies between the Lower 95 % and Upper 95 % values. The probability is 2.5 % that it lies below the lower value, and 2.5 % that it lies above. The narrower this range the better.

## 9.2 Regression Analysis on Data Set 1

Statistical regression analysis are done for the Data set 1 addressed in Chap. 4. Figs. 9.1–9.3 represents Average Response Time (ART) analysis; Figs. 9.4–9.6 represents server utilization analysis; and Figs. 9.7–9.9 represents mean number of calls in the system.

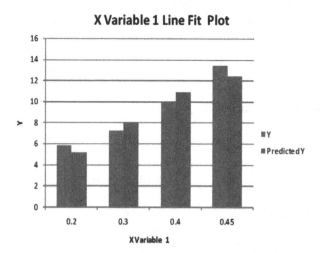

**Fig. 9.1** Experimental data set 1 ART line fit plot

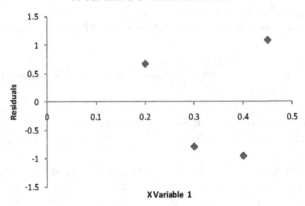

**Fig. 9.2** Experimental data set 1 ART residual plot

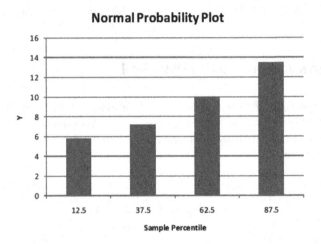

**Fig. 9.3** Experimental data set 1 ART normal probability plot

## 9.2.1   Regression Analysis of ART

```
SUMMARY OUTPUT

Multiple R             0.952654958
R Square               0.907551468
Adjusted R Square      0.861327203
Standard Error         1.257673397
Observations           4
```

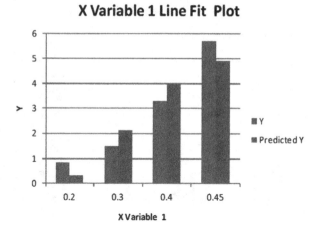

**Fig. 9.4**   Experimental data set 1 MNC line fit plot

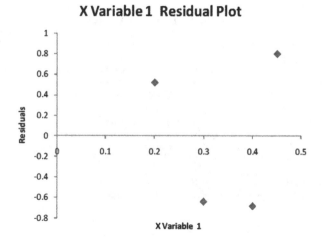

**Fig. 9.5**   Experimental data set 1 MNC residual plot

ANOVA

|  | df | SS | MS |
|---|---|---|---|
| Regression | 1 | 31.05539025 | 31.05539025 |
| Residual | 2 | 3.163484746 | 1.581742373 |
| Total | 3 | 34.218875 | |

| F | Significance F |
|---|---|
| 19.63365892 | 0.047345042 |

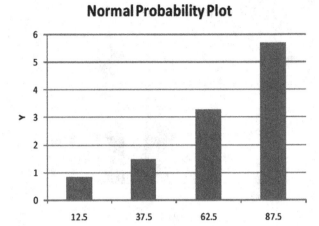

**Fig. 9.6** Experimental data set 1 MNC normal probability plot

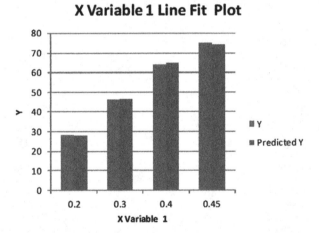

**Fig. 9.7** Experimental data set 1 SU line fit plot

```
                    Coefficients        Standard Error
     Intercept      -0.671864407        2.298132062
     X Variable 1   29.02033898         6.549405196


      t Stat             P-value
  -0.292352393   0.797556104
   4.430988482   0.047345042


  Lower 95%      Upper 95%        Lower 95.0%     Upper 95.0%
  -10.5599286    9.216199782      -10.5599286     9.216199782
  0.840522841    57.20015513      0.840522841     57.20015513
```

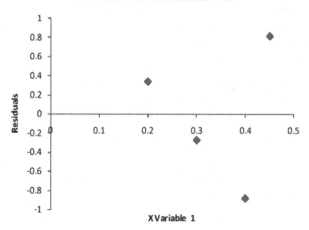

**Fig. 9.8**  Experimental data set 1 SU residual plot

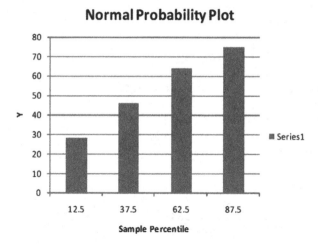

**Fig. 9.9**  Experimental data set 1 SU normal probability plot

```
RESIDUAL OUTPUT

Observation         Predicted Y          Residuals
1                   5.13220339           0.66779661
2                   8.034237288          -0.794237288
3                   10.93627119          -0.956271186
4                   12.38728814          1.082711864
```

```
Standard Residuals
 0.650312295
-0.773442491
 -0.931234003
1.054364199
```

PROBABILITY OUTPUT

| Percentile | Y |
|---|---|
| 12.5 | 5.8 |
| 37.5 | 7.24 |
| 62.5 | 9.98 |
| 87.5 | 13.47 |

## 9.2.2   Regression Analysis of MNC

SUMMARY OUTPUT

Regression Statistics

| Multiple R | 0.934643929 |
|---|---|
| R Square | 0.873559274 |
| Adjusted R Square | 0.81033891 |
| Standard Error | 0.946152781 |
| Observations | 4 |

ANOVA

|  | df | SS | MS |
|---|---|---|---|
| Regression | 1 | 12.36966483 | 12.36966483 |
| Residual | 2 | 1.790410169 | 0.895205085 |

| F | Significance F |
|---|---|
| 13.81768831 | 0.065356071 |

| Total | 3 | 14.160075 |
|---|---|---|

|  | Coefficients | Standard Error |
|---|---|---|
| Intercept | -3.363898305 | 1.728894041 |
| X Variable 1 | 18.31525424 | 4.927144007 |

| t Stat | P-value |
|---|---|
| -1.945693736 | 0.191098538 |
| 3.717215128 | 0.065356071 |

```
Lower 95%            Upper 95%            Lower 95.0%
-10.80272897         4.07493236           -10.80272897
-2.884535371         39.51504385          -2.884535371


Upper 95.0%
4.07493236
39.51504385
```

RESIDUAL OUTPUT

```
Observation Predicted Y        Residuals
1              0.299152542     0.520847458
2              2.130677966     -0.640677966
3              3.96220339      -0.68220339
4              4.877966102     0.802033898
```

Standard Residuals
```
 0.674209563
 -0.829323835
-0.883076306
 1.038190579
```

PROBABILITY OUTPUT

```
Percentile   Y
12.5         0.82
37.5         1.49
62.5         3.28
87.5         5.68
```

## 9.2.3   Regression Analysis of Server Utilization

SUMMARY OUTPUT

Regression Statistics

```
Multiple R          0.999363583
R Square            0.998727571
Adjusted R Square   0.998091356
Standard Error      0.901975234
Observations        4
```

ANOVA

|            | df | SS         | MS         |
|------------|----|------------|------------|
| Regression | 1  | 1277.122881 | 1277.122881 |
| Residual   | 2  | 1.627118644 | 0.813559322 |

| F           | Significance F |
|-------------|----------------|
| 1569.796875 | 0.000636417    |

| Total | 3 | 1278.75 |
|-------|---|---------|

|             | Coefficients | Standard Error | t Stat       |
|-------------|--------------|----------------|--------------|
| Intercept   | -9.559322034 | 1.648168919    | -5.799964992 |
| X Variable 1 | 186.1016949 | 4.697086936    | 39.62066222  |

P-value
 0.028463774
 0.000636417

| Lower 95%     | Upper 95%     | Lower 95.0%   |
|---------------|---------------|---------------|
| -16.65082053  | -2.467823534  | -16.65082053  |
| 165.891761    | 206.3116288   | 165.891761    |

Upper 95.0%
 -2.467823534
 206.3116288

RESIDUAL OUTPUT

| Observation | Predicted Y | Residuals    |
|-------------|-------------|--------------|
| 1           | 27.66101695 | 0.338983051  |
| 2           | 46.27118644 | -0.271186441 |
| 3           | 64.88135593 | -0.881355932 |
| 4           | 74.18644068 | 0.813559322  |

Standard Residuals
0.460287309
-0.368229847
-1.196747003
 1.104689541

```
PROBABILITY OUTPUT

Percentile  Y
12.5        28
37.5        46
62.5        64
87.5        75
```

## 9.3   Regression Analysis on Data Set 2

Statistical regression analysis are done for the Data set 2 addressed in Chap. 6.
Figs. 9.10–9.12 represents Average Response Time (ART) analysis; Figs. 9.13–
9.15 represents mean number of calls in the system analysis; and Figs. 9.16–9.18
represents server utilization analysis.

### 9.3.1   Regression Analysis of ART

```
SUMMARY OUTPUT

Regression Statistics

Multiple R          0.942342933
R Square            0.888010204
```

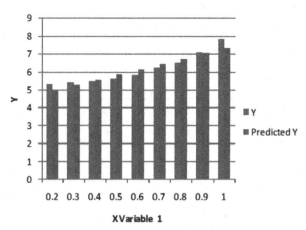

**Fig. 9.10** Experimental data set 2 ART line fit plot

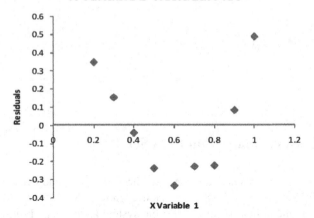

**Fig. 9.11**  Experimental data set 2 ART residual plot

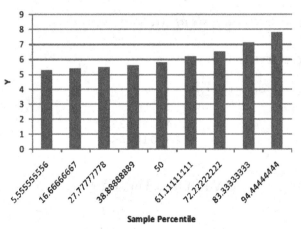

**Fig. 9.12**  Experimental data set 2 ART normal probability plot

```
Adjusted R Square    0.872011662
Standard Error       0.306710659
Observations         9

ANOVA

              df       SS        MS
Regression    1     5.2215    5.2215
Residual      7     0.6585    0.094071429
```

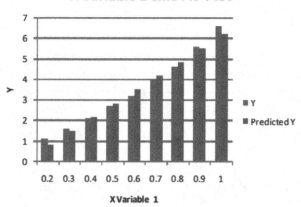

**Fig. 9.13**  Experimental data set 2 MNC line fit plot

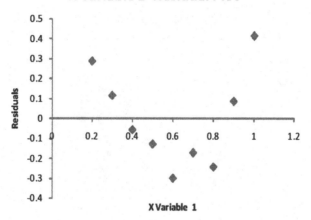

**Fig. 9.14**  Experimental data set 2 MNC residual plot

```
F                Significance F
55.50569476        0.000143193


Total        8    5.88

Coefficients  Standard Error   t Stat          P-value
Intercept     4.363333333      0.258641138     16.8702217
X Variable 1  2.95             0.395961759     7.45021441
```

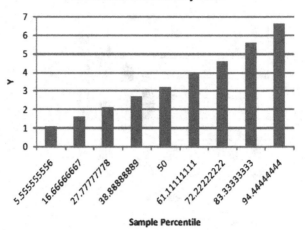

**Fig. 9.15**  Experimental data set 2 MNC normal probability plot

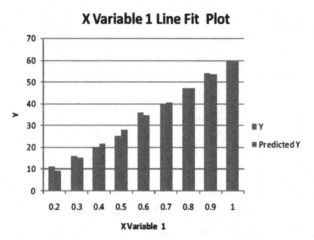

**Fig. 9.16**  Experimental data set 2 SU line fit plot

```
Lower 95%    Upper 95%    Lower 95.0%
3.751744227  4.97492244   3.751744227  4.97492244
2.013699223  3.886300777  2.013699223  3.886300777

Upper 95.0%
6.29639E-07
0.000143193
```

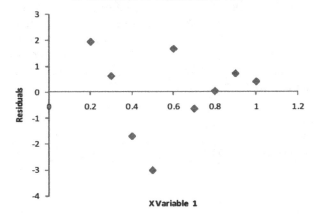

**Fig. 9.17** Experimental data set 2 SU residual plot

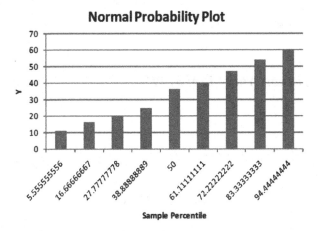

**Fig. 9.18** Experimental data set 2 SU normal probability plot

```
RESIDUAL OUTPUT

Observation Predicted Y  Residuals     Standard
                                       Residuals
1            4.953333333  0.346666667   1.208312278
2            5.248333333  0.151666667   0.528636622
3            5.543333333 -0.043333333  -0.151039035
4            5.838333333 -0.238333333  -0.830714691
```

```
5               6.133333333   -0.333333333  -1.161838729
6               6.428333333   -0.228333333  -0.795859529
7               6.723333333   -0.223333333  -0.778431948
8               7.018333333    0.081666667   0.284650489
9               7.313333333    0.486666667   1.696284544
```

PROBABILITY OUTPUT

```
Percentile      Y
5.555555556     5.3
16.66666667     5.4
27.77777778     5.5
38.88888889     5.6
50              5.8
61.11111111     6.2
72.22222222     6.5
83.33333333     7.1
94.44444444     7.8
```

## 9.3.2   Regression Analysis of MNC

SUMMARY OUTPUT

Regression Statistics

```
Multiple R          0.991396664
R Square            0.982867344
Adjusted R Square   0.980419822
Standard Error      0.259624271
Observations        9
```

ANOVA

|            | df | SS          | MS          |
|------------|----|-------------|-------------|
| Regression | 1  | 27.06816667 | 27.06816667 |
| Residual   | 7  | 0.471833333 | 0.067404762 |
| Total      | 8  | 27.54       |             |

```
 F                 Significance F
401.5764747            1.92856E-07

                  Coefficients    Standard Error
Intercept          -0.53          0.218934408
X Variable 1       6.716666667    0.335173492

t Stat             P-value
-2.420816372       0.046036963
20.03937311        1.92856E-07
```

| Lower 95% | Upper 95% | Lower 95.0% | Upper 95.0% |
|---|---|---|---|
| -1.04769761 | -0.01230238 | -1.04769761 | -0.012302389 |
| 5.924107299 | 7.509226034 | 5.924107299 | 7.509226034 |

RESIDUAL OUTPUT

| Observation | Predicted Y | Residuals | Standard Residuals |
|---|---|---|---|
| 1 | 0.813333333 | 0.286666667 | 1.180396411 |
| 2 | 1.485 | 0.115 | 0.473531119 |
| 3 | 2.156666667 | -0.056666667 | -0.233334174 |
| 4 | 2.828333333 | -0.128333333 | -0.528433277 |
| 5 | 3.5 | -0.3 | -1.23529857 |
| 6 | 4.171666667 | -0.171666667 | -0.706865293 |
| 7 | 4.843333333 | -0.243333333 | -1.001964396 |
| 8 | 5.515 | 0.085 | 0.350001262 |
| 9 | 6.186666667 | 0.413333333 | 1.701966919 |

PROBABILITY OUTPUT

| Percentile | Y |
|---|---|
| 5.555555556 | 1.1 |
| 16.66666667 | 1.6 |
| 27.77777778 | 2.1 |
| 38.88888889 | 2.7 |
| 50 | 3.2 |
| 61.11111111 | 4 |
| 72.22222222 | 4.6 |
| 83.33333333 | 5.6 |
| 94.44444444 | 6.6 |

### 9.3.3   *Regression Analysis of SU*

SUMMARY OUTPUT

Regression Statistics

Multiple R              0.995852349
R Square                0.9917219
Adjusted R Square       0.990539314
Standard Error          1.689604067
Observations            9

ANOVA

|            | df | SS          | MS          |
|------------|----|-------------|-------------|
| Regression | 1  | 2394.016667 | 2394.016667 |
| Residual   | 7  | 19.98333333 | 2.854761905 |
| Total      | 8  | 2414        |             |

| F           | Significance F |
|-------------|----------------|
| 838.6046706 | 1.50693E-08    |

|              | Coefficients | Standard Error |
|--------------|--------------|----------------|
| Intercept    | -3.566666667 | 1.424799253    |
| X Variable 1 | 63.16666667  | 2.181269472    |

| t Stat       | P-value      |
|--------------|--------------|
| -2.503276627 | 0.04079617   |
| 28.95867177  | 1.50693E-08  |

| Lower 95%    | Upper 95%     | Lower 95.0%   |
|--------------|---------------|---------------|
| -6.935781532 | -0.197551801  | -6.935781532  |
| 58.00878398  | 68.32454936   | 58.00878398   |

| Upper 95.0%  |
|--------------|
| -0.197551801 |
| 68.32454936  |

```
RESIDUAL OUTPUT

Observation   Predicted     Residuals     Standard
                                          Residuals
1             9.066666667   1.933333333   1.223257159
2             15.38333333   0.616666667   0.390176852
3             21.7          -1.7          -1.075622674
4             28.01666667   -3.016666667  -1.90870298
5             34.33333333   1.666666667   1.054532033
6             40.65         -0.65         -0.411267493
7             46.96666667   0.033333333   0.021090641
8             53.28333333   0.716666667   0.453448774
9             59.6          0.4           0.253087688


PROBABILITY OUTPUT

Percentile    Y
5.555555556   11
16.66666667   16
27.77777778   20
38.88888889   25
50            36
61.11111111   40
72.22222222   47
83.33333333   54
94.44444444   60
```

## 9.4   Regression Analysis on Data Set 3

Statistical regression analysis are done for Data set 3 addressed in Chap. 7. Figs. 9.19–9.20 represents Call Setup Time (ART) analysis for LAN; Figs. 9.21–9.22 represents queue size analysis for the LAN; Figs. 9.23–9.24 represents memory utilization for the LAN; Figs. 9.25–9.26 represents the CPU utilization for the LAN; Figs. 9.27–9.28 represents Call Setup Time (ART) analysis for WAN; Figs. 9.29–9.30 represents queue size analysis for the WAN; Figs. 9.31–9.32 represents memory utilization for the WAN; and Figs. 9.33–9.34 represents the CPU utilization for the WAN.

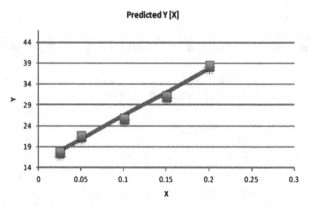

**Fig. 9.19** Experimental data set 3 call setup time LAN fit plot

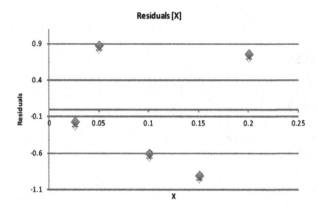

**Fig. 9.20** Experimental data set 3 call setup time residual plot

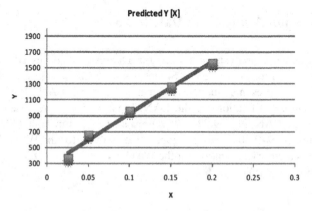

**Fig. 9.21** Experimental data set 3 queue size LAN fit plot

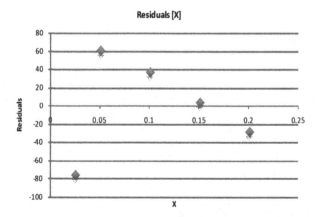

**Fig. 9.22**  Experimental data set 3 queue size residual plot

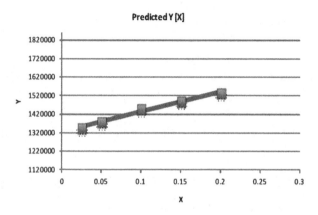

**Fig. 9.23**  Experimental data set 3 memory utilization LAN fit plot

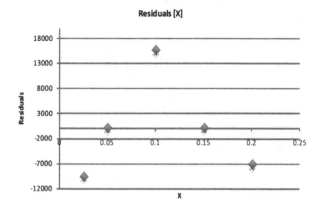

**Fig. 9.24**  Experimental data set 3 memory utilization residual plot

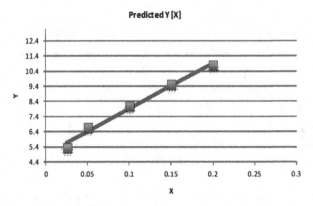

**Fig. 9.25** Experimental data set 3 CPU utilization LAN fit plot

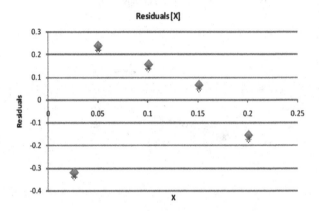

**Fig. 9.26** Experimental data set 3 CPU utilization residual plot

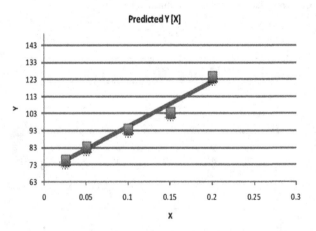

**Fig. 9.27** Experimental data set 3 call setup time WAN fit plot

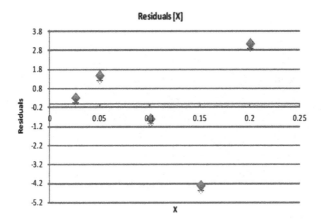

**Fig. 9.28** Experimental data set 3 call setup time residual plot

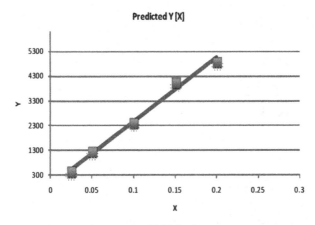

**Fig. 9.29** Experimental data set 3 queue size WAN fit plot

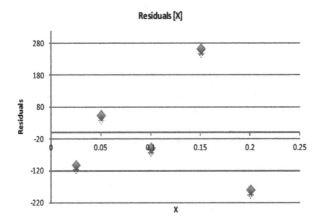

**Fig. 9.30** Experimental data set 3 queue size residual plot

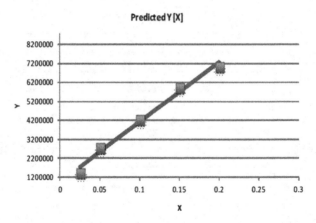

**Fig. 9.31** Experimental data set 3 memory utilization WAN fit plot

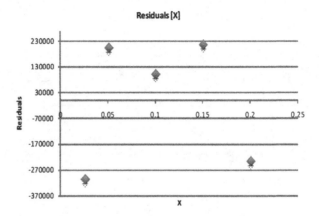

**Fig. 9.32** Experimental data set 3 memory utilization residual plot

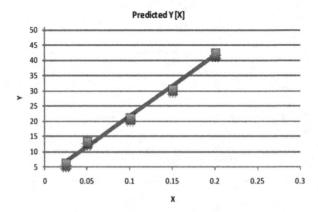

**Fig. 9.33** Experimental data set 3 CPU utilization WAN fit plot

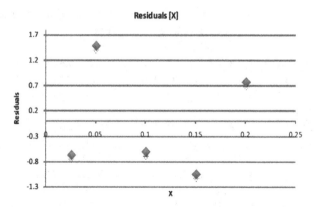

**Fig. 9.34** Experimental data set 3 CPU utilization residual plot

## 9.4.1 Regression Analysis of Call Setup Time (LAN)

```
Linear Regression

Regression Statistics
R                       9.95108E-1
R Square                9.90239E-1
Adjusted R Square       9.86986E-1
Standard Error          9.25248E-1
Total Number Of Cases   5
Y = 15.2449 + 112.7385 * X


ANOVA
                 d.f.      SS                  MS
Regression       1.E+0    2.60555E+2          2.60555E+2
Residual         3.E+0    2.56825E+0          8.56085E-1
Total            4.E+0    2.63123E+2


F                 p-level
3.04356E+2          4.10473E-4


              Coefficients Standard Error  LCL
Intercept     1.52449E+1   7.94747E-1      1.16361E+1
X             1.12739E+2   6.46222E+0      8.33955E+1

  UCL    t Stat
  1.88536E+1   1.9182E+1
  1.42082E+2   1.74458E+1          s
```

```
                         p-level              H0 (2%) rejected?
                         3.09425E-4           Yes
                         4.10473E-4           Yes
        LCL - Lower value of a reliable interval (LCL)
        UCL - Upper value of a reliable interval (UCL)


        Residuals
        Observation       Predicted Y        Residual
        1                 1.80633E+1         -1.63317E-1
        2                 2.08818E+1          8.8822E-1
        3                 2.65187E+1         -5.98707E-1
        4                 3.21556E+1         -8.95634E-1
        5                 3.77926E+1          7.69439E-1


        Standard Residuals
        -2.03818E-1
         1.10849E+0
         -7.47181E-1
         -1.11774E+0
         9.60252E-1
```

## 9.4.2   Regression Analysis of Queue Size (LAN)

```
        Linear Regression

        Regression Statistics
        R                        9.93459E-1
        R Square                 9.86961E-1
        Adjusted R Square        9.82615E-1
        Standard Error           6.23566E+1
        Total Number Of Cases    5
        Y = 272.2927 + 6562.9268 * X

        ANOVA
                         d.f.      SS                  MS
        Regression       1.E+0     8.82976E+5          8.82976E+5
        Residual         3.E+0     1.1665E+4           3.88834E+3
        Total            4.E+0     8.94641E+5


        F                        p-level
        2.27083E+2               6.34383E-4
```

```
            Coefficients Standard Error  LCL
Intercept   2.72293E+2   5.35615E+1      2.90857E+1
X           6.56293E+3   4.35517E+2      4.58537E+3


UCL              t Stat
5.155E+2         5.08374E+0
8.54048E+3       1.50693E+1


                p-level      H0 (2 percent rejected)
                1.47064E-2   Yes
                6.34383E-4   Yes
LCL - Lower value of a reliable interval (LCL)
UCL - Upper value of a reliable interval (UCL)


Residuals
Observation        Predicted Y        Residual
1                  4.36366E+2         -7.53659E+1
2                  6.00439E+2         6.1561E+1
3                  9.28585E+2         3.74146E+1
4                  1.25673E+3         4.26829E+0
5                  1.58488E+3         -2.7878E+1

 Standard Residuals
  -1.3956E+0
  1.13997E+0
  6.92833E-1
  7.9039E-2
  -5.16237E-1
```

### 9.4.3   Regression Analysis of Memory Utilization (LAN)

```
Linear Regression

Regression Statistics

R                       9.91912E-1
R Square                9.83889E-1
Adjusted R Square       9.78519E-1
Standard Error          1.13955E+4
Total Number Of Cases   5
Y = 1327058.1037 + 1077288.5366 * X
```

```
ANOVA
                   d.f.      SS                    MS
Regression         1.E+0     2.37913E+10           2.37913E+10
Residual           3.E+0     3.89575E+8            1.29858E+8
Total              4.E+0     2.41809E+10


     F                     p-level
   1.8321E+2                8.72126E-4


                   Coefficients Standard Error   LCL
Intercept          1.32706E+6   9.78826E+3       1.28261E+6
X                  1.07729E+6   7.95899E+4       7.15895E+5


UCL                   t Stat
1.3715E+6             1.35577E+2
1.43868E+6           1.35355E+1


                   p-level      H0 (2 percent) rejected?
                   8.84774E-7   Yes
                   8.72126E-4   Yes


LCL - Lower value of a reliable interval (LCL)
UCL - Upper value of a reliable interval (UCL)


Residuals
Observation        Predicted Y       Residual
1                  1.35399E+6        -9.35832E+3
2                  1.38092E+6         2.4047E+2
3                  1.43479E+6         1.5892E+4
4                  1.48865E+6         2.48616E+2
5                  1.54252E+6        -7.02281E+3


Standard Residuals
-9.48271E-1
2.43666E-2
1.61033E+0
2.5192E-2
-7.11616E-1
```

## 9.4.4   Regression Analysis of CPU Utilization (LAN)

```
Linear Regression

Regression Statistics
R                       9.94287E-1
R Square                9.88606E-1
Adjusted R Square       9.84808E-1
Standard Error          2.66451E-1
Total Number Of Cases   5
Y = 4.9474 + 30.0244 * X

ANOVA
                  d.f.    SS                 MS
Regression        1.E+0   1.848E+1           1.848E+1
Residual          3.E+0   2.12988E-1         7.09959E-2
Total    4.E+0    1.8693E+1

F                 p-level
2.60297E+2           5.17957E-4

               Coefficients Standard Error LCL
Intercept      4.94744E+0   2.28869E-1     3.90821E+0
X              3.00244E+1   1.86097E+0     2.15743E+1

  UCL                t Stat
  5.98667E+0         2.16169E+1
  3.84745E+1         1.61337E+1

               p-level        H0 (2 percent) rejected?
               2.16649E-4     Yes
               5.17957E-4     Yes

LCL - Lower value of a reliable interval (LCL)
UCL - Upper value of a reliable interval (UCL)

Residuals
Observation         Predicted Y        Residual
1                   5.69805E+0         -3.18049E-1
2                   6.44866E+0         2.41341E-1
3                   7.94988E+0         1.60122E-1
4                   9.4511E+0          6.89024E-2
5                   1.09523E+1         -1.52317E-1

  Standard Residuals
  -1.37831E+0
```

```
   1.04589E+0
 6.93911E-1
 2.98598E-1
-6.60087E-1
```

## 9.4.5   Regression Analysis of Call Setup Time (WAN)

```
Linear Regression

Regression Statistics
R                        9.89133E-1
R Square                 9.78384E-1
Adjusted R Square        9.71179E-1
Standard Error           3.24469E+0
Total Number Of Cases    5
Y = 69.0074 + 264.0741 * X

ANOVA
                 d.f.    SS                 MS
Regression       1.E+0   1.42957E+3         1.42957E+3
Residual         3.E+0   3.1584E+1          1.0528E+1
Total            4.E+0   1.46115E+3


 F               p-level
1.35787E+2       1.35764E-3


              Coefficients Standard Error LCL
Intercept     6.90074E+1   2.78704E+0     5.63523E+1
X             2.64074E+2   2.26619E+1     1.61173E+2


UCL              t Stat
8.16626E+1       2.47601E+1
3.66975E+2       1.16528E+1


                 p-level          H0 (2%) rejected?
                 1.44434E-4       Yes
                 1.35764E-3       Yes


LCL - Lower value of a reliable interval (LCL)
UCL - Upper value of a reliable interval (UCL)
```

```
Residuals
Observation          Predicted Y           Residual
1                    7.56093E+1            3.58732E-1
2                    8.22111E+1            1.51888E+0
3                    9.54148E+1            -7.56829E-1
4                    1.08619E+2            -4.29854E+0
5                    1.21822E+2            3.17776E+0

Standard Residuals
1.27663E-1
5.40529E-1
 -2.69336E-1
 -1.52974E+0
 1.13088E+0
```

## 9.4.6   Regression Analysis of Queue Size (WAN)

```
Linear Regression

Regression Statistics
R                         9.95692E-1
R Square                  9.91403E-1
Adjusted R Square         9.88537E-1
Standard Error            1.98065E+2
Total Number Of Cases     5
Y =- 91.4756 + 25730.2439 * X

ANOVA
                d.f.    SS                    MS
Regression      1.E+0   1.35719E+7            1.35719E+7
Residual        3.E+0   1.17689E+5            3.92297E+4
Total           4.E+0   1.36896E+7


 F                      p-level
3.45961E+2              3.3918E-4


            Coefficients Standard Error   LCL
Intercept   -9.14756E+1  1.70129E+2       -8.63981E+2
X           2.57302E+4   1.38334E+3       1.94489E+4

UCL                  t Stat
 6.81029E+2          5.37684E-1
3.20116E+4           1.86E+1
```

```
                    p-level              H0 (2%) rejected?
                    6.28121E-1           No
                    3.3918E-4            Yes
LCL - Lower value of a reliable interval (LCL)
UCL - Upper value of a reliable interval (UCL)


Residuals
Observation         Predicted Y          Residual
1                   5.5178E+2            -9.87805E+1
2                   1.19504E+3           5.69634E+1
3                   2.48155E+3           -4.55488E+1
4                   3.76806E+3           2.65939E+2
5                   5.05457E+3           -1.78573E+2


Standard Residuals
 -5.75882E-1
  3.32092E-1
 -2.65545E-1
 1.5504E+0
 -1.04107E+0
```

## 9.4.7   Regression Analysis of Memory Utilization (WAN)

```
Linear Regression

Regression Statistics
R                       9.93931E-1
R Square                9.87899E-1
Adjusted R Square       9.83865E-1
Standard Error          2.87968E+5
Total Number Of Cases   5
Y = 978680.9756 + 31475049.7561 * X


ANOVA
                d.f.    SS                  MS
Regression      1.E+0   2.03089E+13         2.03089E+13
Residual        3.E+0   2.48776E+11         8.29254E+10
Total           4.E+0   2.05577E+13


F                       p-level
 2.44906E+2             5.67054E-4
```

```
              Coefficients    Standard Error LCL
 Intercept    9.78681E+5      2.47351E+5      -1.44469E+5
 X            3.1475E+7       2.01125E+6       2.23425E+7

 UCL                  t Stat
 2.10183E+6           3.95664E+0
 4.06076E+7           1.56495E+1

                      p-level           H0 (2%) rejected?
                      2.88178E-2        No
                      5.67054E-4        Yes
```

LCL - Lower value of a reliable interval (LCL)
UCL - Upper value of a reliable interval (UCL)

```
·Residuals
Observation          Predicted Y         Residual
1                    1.76556E+6          -3.02145E+5
2                    2.55243E+6           2.08745E+5
3                    4.12619E+6           1.0527E+5
4                    5.69994E+6           2.20735E+5
5                    7.27369E+6          -2.32604E+5

Standard Residuals
 -1.21155E+0
  8.37029E-1
  4.22115E-1
  8.85107E-1
 -9.32702E-1
```

### 9.4.8   Regression Analysis of CPU Utilization (WAN)

```
Linear Regression

Regression Statistics
R                         9.97096E-1
R Square                  9.94201E-1
Adjusted R Square         9.92268E-1
Standard Error            1.25153E+0
Total Number Of Cases     5
Y = 1.9712 + 198.2366 * X
```

ANOVA

|            | d.f.    | SS          | MS          |
|------------|---------|-------------|-------------|
| Regression | 1.E+0   | 8.05604E+2  | 8.05604E+2  |
| Residual   | 3.E+0   | 4.69897E+0  | 1.56632E+0  |
| Total      | 4.E+0   | 8.10303E+2  |             |

| F          | p-level   |
|------------|-----------|
| 5.14328E+2 | 1.8775E-4 |

|           | Coefficients | Standard Error | LCL         |
|-----------|--------------|----------------|-------------|
| Intercept | 1.97116E+0   | 1.07501E+0     | -2.91013E+0 |
| X         | 1.98237E+2   | 8.74106E+0     | 1.58546E+2  |

| UCL        | t Stat      |
|------------|-------------|
| 6.85245E+0 | 1.83362E+0  |
| 2.37927E+2 | 2.26788E+1  |

| p-level    | H0 (2%) rejected? |
|------------|-------------------|
| 1.64076E-1 | No                |
| 1.8775E-4  | Yes               |

LCL - Lower value of a reliable interval (LCL)
UCL - Upper value of a reliable interval (UCL)
Residuals

| Observation | Predicted Y | Residual    |
|-------------|-------------|-------------|
| 1           | 6.92707E+0  | -6.47073E-1 |
| 2           | 1.1883E+1   | 1.49701E+0  |
| 3           | 2.17948E+1  | -5.94817E-1 |
| 4           | 3.17066E+1  | -1.03665E+0 |
| 5           | 4.16185E+1  | 7.81524E-1  |

Standard Residuals
-5.9701E-1
1.38119E+0
-5.48797E-1
-9.56443E-1
7.21059E-1

## 9.5 Regression Analysis on Data Set 4

Statistical regression analysis are done for the Data set 4 addressed in Chap. 8. Figs. 9.35–9.36 represents Call Setup Time analysis; Figs. 9.37–9.38 represents queue size analysis; Figs. 9.39–9.40 represents memory utilization analysis; and 9.41–9.42 represents the CPU utilization analysis.

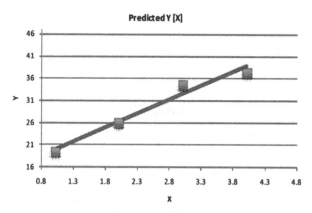

**Fig. 9.35** Experimental data set 4 call setup time LAN fit plot

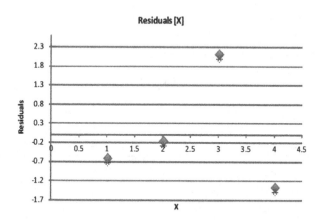

**Fig. 9.36** Experimental data set 4 call setup time residual plot

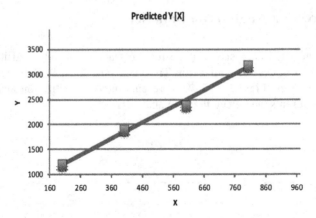

**Fig. 9.37** Experimental data set 4 queue size LAN fit plot

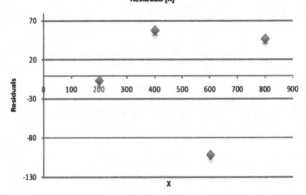

**Fig. 9.38** Experimental data set 4 queue size residual plot

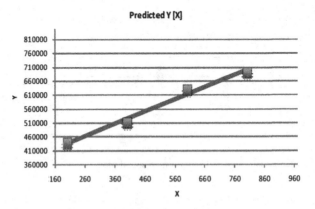

**Fig. 9.39** Experimental data set 4 memory utilization LAN fit plot

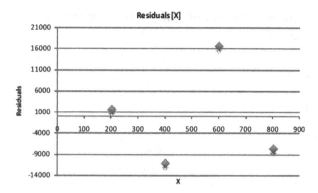

**Fig. 9.40**   Experimental data set 4 memory utilization residual plot

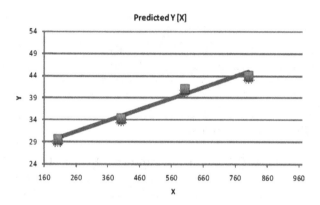

**Fig. 9.41**   Experimental data set 4 CPU utilization LAN fit plot

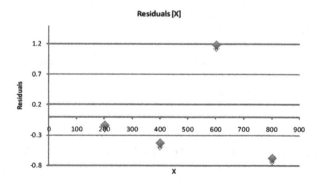

**Fig. 9.42**   Experimental data set 4 CPU utilization residual plot

### 9.5.1  Regression Analysis of Call Setup Time for Secure vs Non-secure

```
Linear Regression

Regression Statistics
R                         9.83355E-1
R Square                  9.66987E-1
Adjusted R Square         9.5048E-1
Standard Error            1.82856E+0
Total Number Of Cases     4
Y = 13.7100 + 6.2590 * X


ANOVA
                d.f.      SS              MS
Regression      1.E+0     1.95875E+2      1.95875E+2
Residual        2.E+0     6.68727E+0      3.34364E+0
Total           3.E+0     2.02563E+2


F                   p-level
5.85816E+1          1.66452E-2


             Coefficients   Standard Error LCL
Intercept    1.371E+1       2.23952E+0     -1.88727E+0
X            6.259E+0       8.17757E-1     5.63683E-1


UCL             t Stat
2.93073E+1       6.12185E+0
1.19543E+1       7.65386E+0


                 p-level          H0 (2%) rejected?
                 2.56604E-2       No
                 1.66452E-2       Yes


LCL - Lower value of a reliable interval (LCL)
UCL - Upper value of a reliable interval (UCL)

Residuals
Observation         Predicted Y      Residual
1                   1.9969E+1        -5.99E-1
2                   2.6228E+1        -1.58E-1
3                   3.2487E+1        2.113E+0
4                   3.8746E+1        -1.356E+0
```

```
Standard Residuals
 -4.01202E-1
 -1.05826E-1
  1.41526E+0
 -9.0823E-1
```

## 9.5.2   *Regression Analysis of Queue Size for Secure vs Non-secure*

```
Linear Regression

Regression Statistics
R                       9.9618E-1
R Square                9.92374E-1
Adjusted R Square       9.88561E-1
Standard Error          8.92054E+1
Total Number Of Cases   4
Y = 564.0000 + 3.2180 * X
```

```
ANOVA
                 d.f.     SS                  MS
Regression       1.E+0    2.0711E+6           2.0711E+6
Residual         2.E+0    1.59152E+4          7.9576E+3
Total            3.E+0    2.08702E+6
```

```
 F                   p-level
 2.60268E+2          3.8202E-3
```

```
            Coefficients  Standard Error LCL
Intercept 5.64E+2         1.09254E+2     -1.96905E+2
X         3.218E+0        1.99469E-1      1.82878E+0
```

```
UCL             t Stat
1.3249E+3        5.16229E+0
4.60722E+0       1.61328E+1
                 p-level          H0 (2%) rejected?
                 3.55364E-2       No
                 3.8202E-3        Yes
```

```
LCL - Lower value of a reliable interval (LCL)
UCL - Upper value of a reliable interval (UCL)
```

```
Residuals
Observation       Predicted Y        Residual
1                 1.2076E+3          -5.6E+0
2                 1.8512E+3          5.88E+1
3                 2.4948E+3          -1.008E+2
4                 3.1384E+3          4.76E+1


Standard Residuals
-7.68852E-2
8.07294E-1
 -1.38393E+0
 6.53524E-1
```

### 9.5.3  Regression Analysis of Memory Utilization for Secure vs Non-secure

```
Linear Regression

Regression Statistics
R                       9.94093E-1
R Square                9.88222E-1
Adjusted R Square       9.82332E-1
Standard Error          1.51937E+4
Total Number Of Cases   4
Y = 350523.0000 + 440.0980 * X


ANOVA
                  d.f.    SS                  MS
Regression        1.E+0   3.87372E+10         3.87372E+10
Residual          2.E+0   4.617E+8            2.3085E+8
Total             3.E+0   3.91989E+10


 F                   p-level
1.67803E+2          5.90663E-3


                Coefficients  Standard Error  LCL
Intercept       3.50523E+5    1.86085E+4      2.20923E+5
X               4.40098E+2    3.39742E+1      2.03482E+2


UCL                      t Stat
4.80123E+5      1.88368E+1
6.76714E+2      1.29539E+1
```

```
                      p-level              HO (2%) rejected?
                      2.80645E-3           Yes
                      5.90663E-3           Yes
```

LCL - Lower value of a reliable interval (LCL)
UCL - Upper value of a reliable interval (UCL)

Residuals

| Observation | Predicted Y | Residual |
|---|---|---|
| 1 | 4.38543E+5 | 1.7404E+3 |
| 2 | 5.26562E+5 | -1.09962E+4 |
| 3 | 6.14582E+5 | 1.67712E+4 |
| 4 | 7.02601E+5 | -7.5154E+3 |

Standard Residuals
```
 1.40291E-1
-8.86387E-1
 1.3519E+0
-6.05805E-1
```

## 9.5.4 Regression Analysis of CPU Utilization for Secure vs Non-secure

Linear Regression

Regression Statistics
| | |
|---|---|
| R | 9.91774E-1 |
| R Square | 9.83616E-1 |
| Adjusted R Square | 9.75424E-1 |
| Standard Error | 1.01971E+0 |
| Total Number Of Cases | 4 |

$Y = 24.9250 + 0.0250 * X$

ANOVA

| | d.f. | SS | MS |
|---|---|---|---|
| Regression | 1.E+0 | 1.2485E+2 | 1.2485E+2 |
| Residual | 2.E+0 | 2.07963E+0 | 1.03982E+0 |
| Total | 3.E+0 | 1.2693E+2 | |

| F | p-level |
|---|---|
| 1.20069E+2 | 8.22589E-3 |

|           | Coefficients | Standard Error | LCL        |
|-----------|--------------|----------------|------------|
| Intercept | 2.4925E+1    | 1.24889E+0     | 1.6227E+1  |
| X         | 2.4985E-2    | 2.28015E-3     | 9.10478E-3 |

| UCL        | t Stat     |
|------------|------------|
| 3.36231E+1 | 1.99577E+1 |
| 4.08652E-2 | 1.09576E+1 |

| p-level    | H0 (2%) rejected? |
|------------|-------------------|
| 2.50118E-3 | Yes               |
| 8.22589E-3 | Yes               |

LCL - Lower value of a reliable interval (LCL)
UCL - Upper value of a reliable interval (UCL)

Residuals

| Observation | Predicted Y | Residual  |
|-------------|-------------|-----------|
| 1           | 2.9922E+1   | -1.22E-1  |
| 2           | 3.4919E+1   | -4.19E-1  |
| 3           | 3.9916E+1   | 1.204E+0  |
| 4           | 4.4913E+1   | -6.63E-1  |

Standard Residuals
 -1.4653E-1
 -5.03247E-1
 1.44609E+0
 -7.96308E-1

# Chapter 10
# Summary and Future Work

This book focused on studying the performance characteristics and scalability aspects of SIP proxy server, which plays a vital role in the SIP architecture based network. Since it is not realistically possible to simulate or emulate the real world SIP network and study the performance characteristics of the SIP proxy server, significant effort was devoted in exploring new creative ways of designing the lab network and hardware setup. We conducted experiments with state of the art equipment, SIP proxy server software and other tools at the high tech CISCO "LASER" lab. We collected several data sets over the course of this research, did the comparative analysis and presented the findings in this thesis. The experimental results presented can be used as a benchmark data for the current and future SIP proxy server implementations. The performance parameters such as average response time or Call setup times, mean number of calls in the system, CPU utilization, Memory utilization, Queue size while establishing the SIP based media sessions is a very fundamental research in IP based telephony. Providing security for the SIP based network is a performance overhead for the service providers. We studied the performance impacts of processing SIP based calls are made using secured and non-secured transport protocols and presented the results.

## 10.1 Summary of Contributions

In the first part of our research (Chap. 4), we studied the $M/M/1$ based performance model of the SIP proxy server provided by authors in Gurbani et al. (2005). We conducted experiments, collected *Data Set 1* and compared that with that model. We proved that their model is not close to the empirical data; we established a better and simpler $M/D/1$ based analytical model with the deterministic service time performed better than the $M/M/1$ model. We did a comparative study of these results and published in Subramanian and Dutta (2008).

S.V. Subramanian and R. Dutta, *Measuring SIP Proxy Server Performance*,
DOI 10.1007/978-3-319-00990-2_10,
© Springer International Publishing Switzerland 2013

In Chap. 5, we surveyed the SIP proxy server architecture, major components, critical functions and importantly the software implementations.

In the second part of our research (discussed in Chap. 6), based on our clear understanding of the software implementation of the SIP proxy server coupled with the latest advancements in hardware and software, we considered the SIP proxy server with the multi threaded mechanisms. Based on the measurements and analysis, we found that the SIP proxy server with $M/M/c$ queuing model produces better prediction of server performance than the $M/M/1$ model with six queuing stations, proposed by authors in Gurbani et al. (2005). Also established that the SIP proxy server architecture modeled by the $M/M/c$ model can scale well in terms of processing number of incoming SIP calls. This provides theoretical justification for preferring the multiple identical threaded realization of the SIP proxy server architecture as opposed to the older, thread-per-module realization. With the analytical and experimental results, we established that the average response time, mean number of calls and server utilization factor of the $M/M/c$ model can produce a more predictable, significant performance improvements and also met the ITU-T standards. This part of research was published in Subramanian and Dutta (2009a).

We expanded our research and investigated the scalability of the single SIP proxy server in the LAN environment as a third part of our research. We measured the key scalability factors such as CPU utilization, memory utilization and queue size and published the results in Subramanian and Dutta (2009b). We discussed this in Sect. 6.6.

We conducted an empirical study and learned the performance impact of Call Hold Time (CHT) in SIP proxy server. The key findings of this empirical study are presented in Sect. 7.1.1. This study initiated the fourth part of our research discussed in Chap. 7. We adopted the same $M/M/c$ that we previously discussed in Chap. 6, enhanced the model slightly (to accommodate multi-proxy setup) and continued our research in evaluating the performance impacts on SPS in LAN and WAN environment with varying CHT. We conducted significant study on emulating the WAN environment to measure the performance characteristics of SIP proxy servers in an IETF standard trapezoidal setup to compare the data when the SIP calls are made in LAN environment. We established that CHT played a significant role in SIP proxy server performance in establishing the SIP sessions. The comparative study of these results were submitted as a paper (Subramanian and Dutta 2010b).

In Chap. 8, as the fifth part of our research, we presented an empirical study and established the performance overheads on the SIP proxy server when the SIP calls are made using UDP, TCP (non-secure transports) and TLS-authentication and TLS-encrypted (secure transports). Again we adopted our established $M/M/c$ based model discussed in Chap. 6 and conducted several experiments in an IETF standard trapezoidal setup. The performance of the SIP proxy server comparisons of all these transport protocols, resulted in a paper (Subramanian and Dutta 2010a).

## 10.2 Future Work

Based on the experimental data during our research, we studied the effort required by the SIP proxy server to parse and process SIP requests and responses of almost identical or very close proximity, since the length of the SIP packet for requests and responses to set up the SIP call is a constant. With the multiple threaded software architecture model, there is a room to study further the $M/D/c$ queuing model with arrival times exponentially distributed, service times with no variance and find whether any room for any enhancement or improvised software architecture model for the SIP proxy server. Because of the nature of the Internet, Internet telephony makes it possible to have many services beyond those possible in traditional telephone networks. SIP based Internet telephony services can extend and generalize existing services for those that create new approaches and architectures for implementing them. Since IP Telephony has a solid growth potential in the telecommunication industries, there are additional research opportunities such as telepresence, video transmissions, which are all SIP based technology and products.

# Chapter 11
# Appendix A

## 11.1   $M/D/1$ **Mathematical Dervivation**

Under the assumption of statistical equilibrium, the traffic intensity of the system $\rho = \lambda/\mu$. The probability distribution of the average response time given in Erlang (1948) and Koba (2000).

$$P(W \leq t) = (1 - \lambda) \times \sum_{k=0}^{T} \frac{[\lambda(k-t)]^k}{k!} \times e^{-\lambda(k-t)} \text{where } t = T + \tau \text{ or } T \leq t \quad (11.1)$$

As given in Sun Microsystems (2003) and Rajagopal and Devetsikiotis (2006), the probability of N customers in the system is given by P(N);

$$P(N) = (1 - \rho) \times \sum_{k=0}^{N} e^{k\rho} \left\{ \frac{(-k\rho)^{N-k}}{(N-k)!} - \frac{(-k\rho)^{N-k-1}}{(N-k-1)!} \right\} \text{where } N \geq 0 \quad (11.2)$$

If the sum the probabilities from 0 to t, we obtain

$$S_t = \sum_{N=0}^{t} = (1 - \rho) \sum_{k=0}^{t} \frac{[\rho(k-t)]^k}{k!} \times e^{-\rho(k-t)} \quad (11.3)$$

Average response time distribution for integral values of t can be written as:

$$P(W \leq t) = P(0) + P(1) + \ldots P(t) \quad (11.4)$$

Average response time distribution for non integral values can be written as: Define

$$P(k, \rho) = \frac{\rho^k}{k!} e^{-\rho} \text{for } k = 0, 1, 2, \ldots. \quad (11.5)$$

S.V. Subramanian and R. Dutta, *Measuring SIP Proxy Server Performance*, DOI 10.1007/978-3-319-00990-2_11, © Springer International Publishing Switzerland 2013

$$P(W \leq T + \tau) = e^{\rho k} \sum_{k=0}^{T} \frac{(-\rho k)^k}{k!} \times P(W \leq T - k) \qquad (11.6)$$

We can simplify Eq. 11.6 as:

$$P(W \leq T + \tau) = \sum_{N=0}^{t} \left\{ P(N) \times \sum_{k=0}^{t-N} P(k, -\rho \tau) \right\} \qquad (11.7)$$

where $PW \leq T - k$ can be calculated from Eq. 11.4.

## 11.2  $M/M/c$ **Mathematical Derivation**

Let arrival rate $\lambda_n = \lambda$, for all n and mean system output rate $= c \times \mu$;

$$\mu_n = n \times \mu \text{ and if } 1 \leq n \leq c$$

$$c \times \mu \text{ and if } n \geq c$$

The original birth-death systems uses the equations as shown in

$$p_n = p_0 \prod_{i=1}^{n} \frac{\lambda_{i-1}}{\mu_i} = p_0 \left(\frac{\lambda}{\mu}\right)^n \left(\frac{1}{n!}\right) if 1 \leq n \leq c \qquad (11.8)$$

$$= \prod_{i=1}^{c} \frac{\lambda}{\mu_i \mu} \prod_{i=c+1}^{n} \frac{\lambda}{c\mu} \qquad (11.9)$$

$$= p_0 \left(\frac{\lambda}{\mu}\right)^n \frac{1}{c!} \left(\frac{1}{c}\right)^{n-c} if n \geq c \qquad (11.10)$$

$$p_n = p_0 \prod_{i=1}^{n} \frac{\lambda_{i-1}}{\mu_i} = p_0 \left(\frac{\lambda}{\mu}\right)^n \left(\frac{1}{n!}\right) \text{ if } 1 \leq n \leq c \dots\dots\dots\dots\dots\dots (1)$$

$$= \prod_{i=1}^{c} \frac{\lambda}{\mu_i \mu} \prod_{i=c+1}^{n} \frac{\lambda}{c\mu} = p_0 \left(\frac{\lambda}{\mu}\right)^n \frac{1}{c!} (\frac{1}{c})^{n-c} \text{ if } n \geq c \dots\dots\dots\dots\dots (2)$$

Let $\rho = \frac{\lambda}{c\mu} < 1$ and $c\rho = \frac{\lambda}{\mu}$;
    Mean arrival rate must be less than the mean maximum potential service rate of the system.

$$p_n = p_0 \frac{(c\rho)^n}{n!} = p_0 \frac{p^n c^c}{c!} = \frac{(\rho c)^n}{c^{n-c} c!} for n \geq c \qquad (11.11)$$

Solve for $p_0$ :

$$\sum_{n=0}^{\alpha} p_n = p_0 + \sum_{n=1}^{\alpha} p_n = p_0 \left[ 1 + \sum_{n=1}^{c-1} \frac{(c\rho)^n}{n} + \frac{(c\rho)^c}{c!} \frac{1}{1-\rho} \right]^{-1} \tag{11.12}$$

Queue Length:

$$L_q = \sum_{n=c}^{\alpha} (n-c) p_n \, with \, p_n = \frac{(\rho c)^n}{c^{n-c} c!} \tag{11.13}$$

$$L_q = \sum_{n=c}^{\alpha} \frac{n}{c^{n-c} c!} (\rho c)^n p_0 - \sum_{n=c}^{\alpha} \frac{c}{c^{n-c} c!} (\rho c)^n p_0 \tag{11.14}$$

By considering the two terms on the R.H.S separately, we can derive this equation.

$$\sum_{n=c}^{\alpha} \frac{n(\rho c)^n}{c^{n-c}} \frac{p_0}{c!} = \frac{p_0}{c!} \left[ \frac{(\rho c)^{c+1}}{c} \right] \sum_{n=c}^{\alpha} \left[ (n-c)\rho^{n-c-1} + c\rho^{n-c-1} \right] \tag{11.15}$$

Using derivatives of geometric series,

$$= p_0 \div c! \left[ \frac{(\rho c^{c+1})}{c} \right] \left[ \frac{1}{(1-\rho)^2} + \frac{\frac{c}{\rho}}{1-\rho} \right] \tag{11.16}$$

Now for the second term, we have,

$$\sum_{n=c}^{\alpha} \frac{c(\rho c)^n}{c^{n-c}} \frac{p_0}{c!} = \sum c\rho^c \rho^{n-c} c^c \frac{p_0}{c!} = c(\rho c)^c \frac{p_0}{c!} \sum \rho^{n-c}$$

$$= \frac{p_0 c (\rho c)^c}{c! (1-\rho)} = \left[ \frac{(\rho c)^{c+1}}{c} \right] \frac{\frac{c}{\rho}}{1-\rho} \frac{p_0}{c!} \tag{11.17}$$

Therefore

$$L_q = \frac{(\rho c)^{\frac{c+1}{c}}}{c!(1-\rho)^2} p_0 = \frac{(\frac{\lambda}{\mu})^c \lambda \mu}{(c-1)!(c\mu - \lambda)^2} p_0 \tag{11.18}$$

$L_q = \lambda W_q$ to find $W_q$;
$W = W_q + \frac{1}{\mu}$ to find $W$;
$L = \lambda W$ to find $L$;

$$W_q = \left[ \frac{\frac{\lambda}{\mu}^c \mu}{(c-1)!(c\mu - \lambda)^2} \right] p_0 \tag{11.19}$$

$$W = \frac{1}{\mu} + \left[ \frac{(\frac{\lambda}{\mu})^c \mu}{(c-1)!(c\mu - \lambda)^2} \right] p_0 \tag{11.20}$$

$$L = \frac{\lambda}{\mu} + \left[ \frac{(\frac{\lambda}{\mu})^c \lambda \mu}{(c-1)!(c\mu - \lambda)^2} \right] p_0 \tag{11.21}$$

# Chapter 12
# Appendix B

## 12.1 Lab Hardware Setup

### 12.1.1 SIP Proxy Server Hardware Rack

Refer to Figs. 12.1 and 12.2.

### 12.1.2 SIP Proxy Server Network Diagram

Refer to Fig. 12.3.

### 12.1.3 SIP Proxy Servers in LAN/WAN Setup

Refer to Fig. 12.4.

S.V. Subramanian and R. Dutta, *Measuring SIP Proxy Server Performance*,
DOI 10.1007/978-3-319-00990-2_12,
© Springer International Publishing Switzerland 2013

**Fig. 12.1** SIP proxy server
lab hardware rack with SIP-P

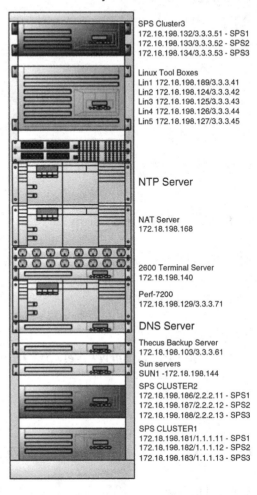

**SIP Proxy Server Rack**

SPS Cluster3
172.18.198.132/3.3.3.51 - SPS1
172.18.198.133/3.3.3.52 - SPS2
172.18.198.134/3.3.3.53 - SPS3

Linux Tool Boxes
Lin1 172.18.198.189/3.3.3.41
Lin2 172.18.198.124/3.3.3.42
Lin3 172.18.198.125/3.3.3.43
Lin4 172.18.198.126/3.3.3.44
Lin5 172.18.198.127/3.3.3.45

NTP Server

NAT Server
172.18.198.168

2600 Terminal Server
172.18.198.140

Perf-7200
172.18.198.129/3.3.3.71

DNS Server

Thecus Backup Server
172.18.198.103/3.3.3.61

Sun servers
SUN1 -172.18.198.144

SPS CLUSTER2
172.18.198.186/2.2.2.11 - SPS1
172.18.198.187/2.2.2.12 - SPS2
172.18.198.188/2.2.2.13 - SPS3

SPS CLUSTER1
172.18.198.181/1.1.1.11 - SPS1
172.18.198.182/1.1.1.12 - SPS2
172.18.198.183/1.1.1.13 - SPS3

**SIP Proxy Server Rack**

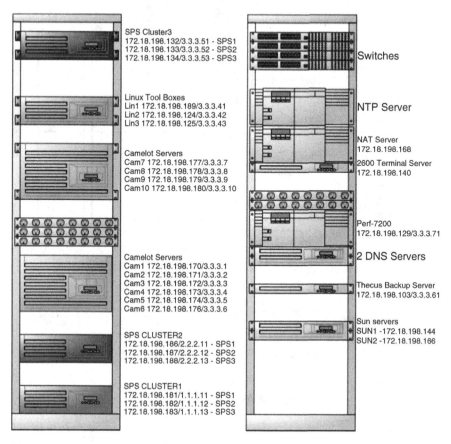

**Fig. 12.2**  SIP proxy server lab hardware rack

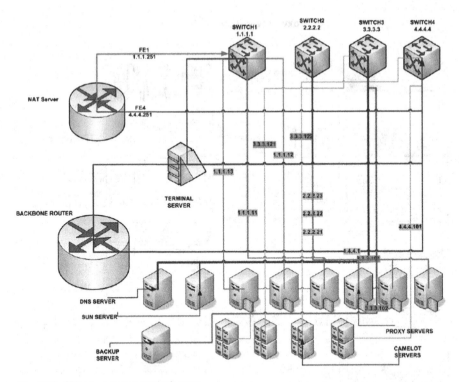

**Fig. 12.3** SIP proxy server network setup

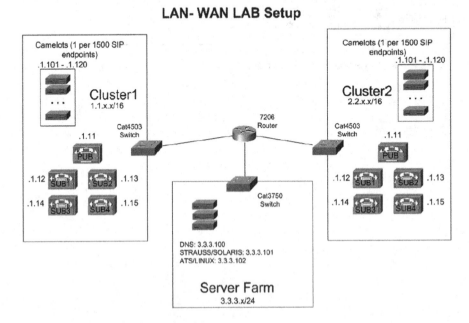

**Fig. 12.4** LAB setup for data set 3 and 4

# Chapter 13
# Appendix C

## 13.1 Tcl/Tk Source Code

### 13.1.1 Inserting the SIP Phones/Devices

```
#!/bin/sh
# \
exec $AUTOTEST/bin/expect "$0" "$@"$
###############################################
#
# File: ccm_add_phones_with_MUSTER_phoneinserter.exp
#
# Author: Sureshkumar V. Subramanian
###############################################
set config_file "/opt/modela/Suresh/scalnewfiles/
Scal-cls3-cls4_decoded_passphrases_02072008.cfg"

source $config_file
# check that we have just one command line
if { [ llength $argv ] < 8 } {
  puts "\nERROR: incorrect number of command line
  arguments!
  Expecting 7, but got only $argc"
  puts "\nUSAGE:
  ./ccm_add_phones_with_phoneinserter.exp
  <CCM_Name>
  <MAC_prefix> <numberOfPhones>
  <startDN> <devicePool> <phoneButtonTem
plate> <numberOfLines> <securityProfile>"
  return
```

```tcl
}
set ccm_name              [ lindex $argv 0 ]
set MAC_prefix            [ lindex $argv 1 ]
set numberOfPhones        [ lindex $argv 2 ]
set startDN               [ lindex $argv 3 ]
set devicePool            [ lindex $argv 4 ]
set phoneButtonTemplate   [ lindex $argv 5 ]
set numberOfLines         [ lindex $argv 6 ]
set securityProfile       [ lindex $argv 7 ]
if { ![ info exists cfg_ccm($ccm_name) ] } {
  puts "\nERROR: CCM NAME '$ccm_name' was NOT found
   in the config file
    '$config_file'"
  return
}
puts "\nCollected command line arguments:"
puts "  - ccm_name:               $ccm_name"
puts "  - MAC_prefix:             $MAC_prefix"
puts "  - numberOfPhones:         $numberOfPhones"
puts "  - startDN:                $startDN"
puts "  - devicePool:             $devicePool"
puts "  - phoneButtonTemplate:
$phoneButtonTemplate"
puts "  - numberOfLines:          $numberOfLines"
puts "  - securityProfile:        $securityProfile"
if { [ regexp {^[0-9]+$} $numberOfLines ] } {
  if { [ expr $numberOfLines < 1 ] ||
       [ expr $numberOfLines > 20 ] } {
    puts "\nERROR: numberOfLines is
    '$numberOfLines'.
    The Lines per Phone must be between 1 and 20."
    puts "SCRIPT EXECUTION STATUS: ABORTED"
    return
  }
} else {
  # $numberOfLines must not be a numeric value
  puts "\nERROR: numberOfLines is
   '$numberOfLines'.
  Must be a numeric value between 1 and 20."
  puts "SCRIPT EXECUTION STATUS: ABORTED"
  return
}
set env(TERM) "vt100"
puts "Before spawning ssh, setting the current
environment variable TERM to $env(TERM)"
```

```
set info_needed [ split $cfg_ccm($ccm_name) "," ]
set ipaddr [ lindex $info_needed 0 ]
set user_name [ lindex $info_needed 1 ]
set passwd [ lindex $info_needed end ]
set prompt "(%|#|\\\$) $"          ; # default prompt
catch  { set prompt $env(EXPECT_PROMPT)}
set timeout -1
eval spawn ssh -l $user_name $ipaddr
set error_counter 0
set my_hostname ""
expect {
  -re "Are you sure you want to continue connecting
  (yes/no)?"
  {send "yes\r" ; exp_continue}
  -re "password:" {
      send "$passwd\r"
      exp_continue
      }
  -re $prompt {
        # check the local env variables
        send "env\r"
        expect -re $prompt
        send "ls -l /usr/local/cm/bin/PhoneInserter.
        class\r"
        expect -re $prompt
        if { [ regexp {No such file or directory}
        $expect_out(buffer)" ] } {
          puts "ERROR: failed to locate the
          phoninserter
          binary in /usr/local/cm/bin:"
          puts "\n######## Expect_out(buffer) is:\n'
          $expect_out(buffer)'\n"
          puts "\nAborting script."
          return
        }
        send "uname -a\r"
        expect -re $prompt
        if { [ regexp -nocase {Linux ([a-z0-9_\-\.]+)
        (.*)}
        $expect_out(buffer) trash my_hostname
        ] } {
          puts "Extracted hostname '$my_hostname'"
        } else {
          puts "Failed to extract the hostname. Using
          '$ccm_name'
```

```
    as CCM hostname"
    set my_hostname $ccm_name
}
set echo_TMOUT_cmd {echo $TMOUT}
send "$echo_TMOUT_cmd\r"
expect -re $prompt
set TMOUT_cmd {export TMOUT="18000"}
send "$TMOUT_cmd\r"
expect -re $prompt
send "$echo_TMOUT_cmd\r"
expect -re $prompt
set dblcn_cmd {dbnotifymon | grep DB}
puts "\nBEFORE running PI, checking the size
of Change Notification
DB queue table: dblcnqueue"
send "$dblcn_cmd\r"
expect -re $prompt
send "cd /usr/local/cm/bin\r"
expect -re $prompt
set pi_cmd {java PhoneInserter -action=insert}
append pi_cmd " -prefix=$MAC_prefix
  -phonecount
=$numberOfPhones
-startingdn=$startDN
-devicepool=$devicePool-linecount=$number
  OfLines
-phonetemplate=\"$phoneButtonTemplate\"
-securityProfile=\"$securityProfile\" -cn
-phonespertx=0"
puts "\nGetting ready to execute the
  Zphoneinserter
  command: $pi_cmd\n"
send "$pi_cmd\r"
expect -re $prompt
log_user 1
puts "\n######## Expect_out(buffer) is:\n
'$expect_out(buffer)'\n"
if {![ regexp {Done inserting phones, ([0-9]+)
phones added in ([0-9\.]+)ms}
$expect_out(buffer) trash num_eps num_seconds]}
{
  puts "ERROR: zPhoneinserter failed to add
    endpoints:"
  puts "\n######## Expect_out(buffer) is:\n
  '$expect_out(buffer)'\n"
```

```
            incr error_counter
        } else {
          if { [ string match $num_eps $numberOfPhones
            ] } {puts "\nzPhoneinserter successfully
                inserted
            $num_eps endpoints in
            $num_seconds
            milliseconds."
          } else {
            puts "ERROR: Phoneinserter inserted ONLY
            $num_eps out of
            $numberOfPhones phones"
            incr error_counter
          }
        }
        sleep 5
        puts "\nAFTER running the PI command,
        checking the size of
        Change Notification
        DB queue table: dblcnqueue"
        send "$dblcn_cmd\r"
        expect -re $prompt
        set prompt {$}
        send "su - informix\r"
        expect -re $prompt
        send "installdb -u\r"
        expect -re $prompt
        log_user 1
        puts "\n######## Expect_out(buffer) is:\n
        '$expect_out(buffer)'\n"
        set prompt {#}
    send "exit\r"
        expect -re $prompt
        if { ![ regexp {Update Statistics(.*)\n}
        $expect_out(buffer) ] } {
          puts "ERROR: Failed to successfully update
          statistics"
          puts "\n######## Expect_out(buffer) is:\n
          '$expect_out(buffer)'\n"
          # we treat this as a major error if unable
          to update the statistics successfully
          incr error_counter
        } else {
          puts "\nSuccessfully updated statistics"
        }
```

```
      puts "Done ..."
      # 10/04/06: report PASS/FAIL status
      ####puts "ERROR_COUNTER IS
      $error_counter"
      if { [string match $error_counter "0" ] } {
        puts "SCRIPT EXECUTION STATUS: PASS"
      } else {
        puts "SCRIPT EXECUTION STATUS: FAIL"
      }
      # exit the ssh session
      send "exit\r"
    }
```

### 13.1.2   Perfmon Source Code

```
#!/bin/sh
# \
exec $AUTOTEST/bin/expect "$0" "$@"
####################################
#
# File: perfmon.tcl
#
# Author: Sureshkumar V. Subramanian
#################################
set start_time [ clock seconds ]
puts "Script [ info script ] $argv started at
 timestamp:
$start_time,
[ clock format $start_time]"
source  "/opt/modela/Suresh/scalnewfiles/Perfmon/
SOAP_Perfmon_lib_021407.exp"
source  "/opt/modela/Suresh/scalnewfiles/Perfmon
/config_perfmon
_SCAL_CLS1_3_CCMs_111907_Perfmon_dir.cfg"
# output the process ID of this script instance
set script_pid [ pid ]
puts "\n\nProcess ID of this script is: $script_pid"
set loop_break 0
set debug_flag 0
set SOAP_error ""
set SOAP_rc 1
set max_reopen_session_attempts 1000
set min_failed_collections 5
```

```
set file_created_rc 0
set file_writing_errors 0
set execution_status "PASS"
trap {
  set trap_name [trap -name]
  set ts [ clock seconds]
  set fts [ clock format $ts]
  puts "\n\nWARNING: got signal named
   '$trap_name' on $fts"
  puts "\nSetting a flag to stop collecting perfmon
  counter values,
  breaks the for loop"
  puts "\nSetting loop_break flag to 1, so no new
  rounds of perfmon
  collections will be made"
  set loop_break 1
} {HUP INT QUIT PIPE TERM CONT WINCH USR1 USR2}
puts "ARGV is: '$argv', ARGC is: '$argc'"
puts "lindex argv 0 is: [ lindex $argv 0 ]"
if { $argc < 1 } {
  append error_message "\nERROR: expecting at
   least one
  command line argument,
  the config file to source"
  append error_message "\nARGC: $argc"
  append error_message "\nARGV: $argv"
  append error_message "\nAborting the script"
  puts $error_message
  puts "SCRIPT EXECUTION STATUS: ABORTED"
  return
}
set cfg_file [ lindex $argv 0 ]
puts "Sourcing the config file passed in the first
command line
argument: '$cfg_file'"
if { [ catch { source $cfg_file } errmsg ] } {
  append error_message "\nERROR: in sourcing
  the config
  file '$cfg_file': $errmsg"
  append error_message "\nAborting the script"
  puts $error_message
  puts "SCRIPT EXECUTION STATUS: ABORTED"
  return
}
puts "\nSuccessfully sourced the config file
```

```
  '$cfg_file'"
puts "\nValidating the config file parameters"
puts "\nCheck that we have the CCM_info(
CCM_group_size)
variable defined in config file"
if { ![ info exist CCM_info(CCM_group_size) ] ||
       $CCM_info(CCM_group_size) == 0 } {
  append error_message "\nERROR: variable
  CCM_info
  (CCM_group_size)
  is not defined or is set to 0"
  append error_message "\nThis variable should
   be defined and must
  be greater than or equal to 1"
  append error_message "\nAborting the script"
  puts $error_message
  puts "SCRIPT EXECUTION STATUS: ABORTED"
  return
}
# Check that we have at least 3 arrays defined in
config file
puts "\nCheck that we have at least 3 arrays defined
in config file"
set req_array_list "CCM_info
perfmon_reference_counters
CCM_skip_counters"
set req_array_list [ join $req_array_list ]
set missing_req_array_counter 0
set missing_req_array_list {}
foreach req_array $req_array_list {
  if { ![ info exist $req_array ] } {
    incr missing_req_array_counter
    lappend missing_req_array_list $req_array
  }
}
if { $missing_req_array_counter > 0 } {
  append error_message "\nERROR: config file does not
  have all 3 arrays defined"
  append error_message "\nMissing
  $missing_req_array_counter required arrays:
  $missing_req_array_list"
  append error_message "\nRequired arrays in the
  config file:"
  append error_message "\n- CCM_info()"
  append error_message
```

```
    "\n-perfmon_reference_counters()"
  append error_message "\n- CCM_skip_counters()"
  append error_message "\nAborting the script"
  puts $error_message
  puts "SCRIPT EXECUTION STATUS:
  ABORTED"
  return
}
puts "\nCheck if all of the required script
CCM_info variables
 are defined and
not set to null in config file"
set req_script_ccm_info_list "debug_flag
nat_flag cert
key iterations
polling_interval temp_result_file_path"
set req_script_ccm_info_list [ join
$req_script_ccm_info_list ]
set missing_script_ccm_info_param_counter 0
set missing_script_ccm_info_array_elements_list {}
foreach req_script_param $req_script_ccm_info_list {
  # check if the config file parameter is missing, or
  set to null
  if { ![ info exist CCM_info($req_script_param) ] ||
       [ string match $CCM_info
       ($req_script_param) "" ] } {
    append missing_script_ccm_info_array_elements_list
    "- CCM_info($req_script_param)" "\n"
    incr missing_script_ccm_info_param_counter
  }
}
if { $missing_script_ccm_info_param_counter > 0 } {
  append error_message "\nERROR: missing
  $missing_script_ccm_info_param_counter
  required script CCM_info parameters:"
  append error_message "\n$missing_script_ccm_info_
  array_elements_list"
  append error_message "\nRequired script CCM_info
  parameters are:"
  append error_message "\n - CCM_info(debug_flag)"
  append error_message "\n - CCM_info(nat_flag)"
  append error_message "\n - CCM_info(cert)"
  append error_message "\n - CCM_info(key)"
  append error_message "\n - CCM_info(iterations)"
  append error_message "\n - CCM_info
```

```tcl
      (polling_interval)"
    append error_message "\nAborting the script"
    puts $error_message
    puts "\nContents of the CCM_info() array in the
    config file:\n"
    foreach elem [ lsort [ array names CCM_info ] ] {
      puts "CCM_info($elem) = $CCM_info($elem)"
    }
    puts "SCRIPT EXECUTION STATUS: ABORTED"
    return
}
set fail_check_flag 0
if { $CCM_info(nat_flag) == 1 } {
  puts "\nNAT Flag is set to 1, checking if
  INSIDE ip addresses are
  defined in the config file ..."
  # check that we have all the inside addresses
    defined
  for { set i 1 } { $i <= $CCM_info(CCM_group_size) }
   { incr i } {
    if { ![ info exists CCM_info
    ($i,CCM_inside_ipaddr) ] ||
        [ string match $CCM_info
        ($i,CCM_inside_ipaddr) "" ] } {
      append error_message "\n
      CONFIG ERROR: failed to
      define the
      inside ip address for CCM $i"
      incr fail_check_flag
    }
  }
  if { $fail_check_flag > 0 } {
    append error_message "\nERROR: failed to
    define all
    he the inside
    ip addresses needed for NAT traversal"
    append error_message "\nAborting script ..."
    puts $error_message
    puts "SCRIPT EXECUTION STATUS: ABORTED"
    return
  } else {
    puts "\nAll INSIDE ip addresses are defined in the
    config file, will be using these
    addresses to prepend perfmon counter names"
  }
```

```
}
# we made it past the basic config file error check
puts "\nCheck that we have at least one
perfmon counter
in perfmon reference array"
if { [ array size perfmon_reference_counters ]
== 0 } {
  append error_message "\nERROR:
  perfmon_reference_counters() array is empty"
  append error_message "\nThis array must contain
  at least one counter"
  append error_message "\nAborting the script"
  puts $error_message
  puts "SCRIPT EXECUTION STATUS:
  ABORTED"
  return
}
puts "\nCheck that all of the required
CCM related
parameters in CCM_info array
are defined for each CCM in the
CCM_group_size"
set req_ccm_info_list
"CCM_Platform SOAP_Protocol
CCM_IP CCM_Username
CCM_Passwd CCM_Name"
set req_ccm_info_list [ join $req_ccm_info_list ]
set missing_ccm_info_param_counter 0
set missing_ccm_info_array_elements_list {}
for { set z 1 } { $z <= $CCM_info
(CCM_group_size) } { incr z } {
  foreach req_elem $req_ccm_info_list {
  # check if the config file parameter is missing,
  or set to null
    if { ![ info exist CCM_info($z,$req_elem) ] ||
      [ string match $CCM_info($z,
      $req_elem) "" ] } {
      append missing_ccm_info_array_elements_list " -
      CCM_info($z" "," "$req_elem" ")" "\n"
      incr missing_ccm_info_param_counter
    }
  }
}
if { $missing_ccm_info_param_counter > 0 } {
  append error_message "\nERROR: Missing
```

```
  $missing_ccm_info_param_counter
  required CCM_info parameters:"
  append error_message "\n
  $missing_ccm_info_array_elements_list"
  append error_message "\n
  Required CCM_info parameters for each CCM
  in the CCM_group_size:"
  append error_message "\n
   - CCM_info(<ccm_index>,CCM_Platform)"
  append error_message "\n
    - CCM_info(<ccm_index>,SOAP_Protocol)"
  append error_message "\n
  - CCM_info(<ccm_index>,CCM_IP)"
  append error_message "\n
  - CCM_info(<ccm_index>,CCM_Username)"
  append error_message "\n
  - CCM_info(<ccm_index>,CCM_Passwd)"
  append error_message "\n
    - CCM_info(<ccm_index>,CCM_Name)"
  append error_message "\nAborting the script"
  puts $error_message
  puts "\n
   Contents of the CCM_info() array in the
   config file:\n"
 foreach elem [ lsort [
 array names CCM_info ] ] {
    puts "CCM_info($elem) = $CCM_info($elem)"
  }

  puts "SCRIPT EXECUTION STATUS: ABORTED"
  return
}
puts "\nCheck that we have a
CCM_skip_counters() array element
for each CCM
in the CCM_group_size"
set skip_counter_array_check_counter 0
set missing_skip_array_elements_list {}
for { set z 1 } { $z <=
$CCM_info(CCM_group_size) } { incr z } {
  if { ![ info exist
  CCM_skip_counters($z) ] } {
    incr skip_counter_array_check_counter
    lappend missing_skip_array_elements_list $z
  }
```

```
}
if { $skip_counter_array_check_counter > 0 } {
  append error_message "\nERROR: the config
  file does not contain a
  CCM_skip_counters()
  array element for each CCM in the CCM_group_size
 $CCM_info(CCM_group_size)"
  append error_message "\nMissing
  $skip_counter_array_check_counter
  CCM array indices:
  $missing_skip_array_elements_list"
  append error_message "\nAborting the script"
  puts $error_message
  puts "SCRIPT EXECUTION STATUS: ABORTED"
  return
}
puts "\nSurvived basic error checking in config file"
puts "\nOutputing contents of the CCM_info() array:\n"
foreach elem [ lsort [ array names CCM_info ] ] {
 puts "CCM_info($elem) = $CCM_info($elem)"
}
puts "\nReference Perfmon Counters in the
sourced config file:"
set total_ref_counters
[ array size perfmon_reference_counters ]
puts "Total number of perfmon counters in the
reference array
perfmon_reference_counters:
$total_ref_counters"
for { set i 0 } { $i <
$total_ref_counters } { incr i } {
  puts "perfmon_reference_counters($i) =
  $perfmon_reference_counters($i)"
}
puts "\nSkipped Perfmon Counters for each CCM:\n"
for { set i 1 } { $i <= $CCM_info(CCM_group_size) }
{ incr i } {
  puts "For CCM $i, $CCM_info($i,CCM_Name) -
  $CCM_info($i,CCM_IP):
  CCM_skip_counters($i) = '$CCM_skip_counters($i)'"
  puts "Skipped counter names:"
  if { [ string match $CCM_skip_counters($i) "" ] } {
    puts "No counters are skipped"
  } else {
    foreach index $CCM_skip_counters($i) {
```

```
        puts "$perfmon_reference_counters($index)"
      }
    }
    puts "\n"
}
if { [ info exists CCM_info(debug_flag) ] } {
  set debug_flag $CCM_info(debug_flag)
  puts "\nSetting the debug flag to: $debug_flag"
}
puts "\nTotal CCMs to collect perfmon counter
values from:
$CCM_info(CCM_group_size)"
# array names returns a list
set reference_counters_indices
 array names perfmon_reference_counters ]
# sort the list in numerical order
set reference_counters_indices
[ lsort -integer $reference_counters_indices ]
puts "\nList of indices of
perfmon_reference_counters array: \n
$reference_counters_indices"
for { set i 1 } { $i <=
$CCM_info(CCM_group_size) } { incr i } {
  # initialize variable
  set CCM_info($i,perfmonSessionHandle) ""
  set CCM_info($i,pingCheck) ""
  set CCM_info($i,failedCollections) 0
  set CCM_info($i,sessionReOpenAttempts) 0
set skip_counters_list [ join
$CCM_skip_counters($i) ]
  set skip_counters_list_len [ llength
  $skip_counters_list ]
  puts "\nCCM $i: list of indices of counters to skip:
  $skip_counters_list. Its length is:
  $skip_counters_list_len"
  set CCM_info($i,total_counters) [ expr
  $total_ref_counters -
  $skip_counters_list_len ]
  set CCM_info($i,skipped_counter_indices)
  $CCM_skip_counters($i)
  # initialize the list to null
  set CCM_info($i,collected_counter_indices) {}
  # check if no counters are skipped
  if { $skip_counters_list_len == 0 } {
    set CCM_info($i,collected_counter_indices) [ join
```

```
        $reference_counters_indices ]
        puts "\nFor CCM $i: list of indices of counters
        to collect is:\n
        $CCM_info($i,
        collected_counter_indices)"
      } else {
        foreach ref_counter $reference_counters_indices {
          set found_skip 0
          foreach skip_counter $skip_counters_list {
            if { $ref_counter == $skip_counter } {
              set found_skip 1
              # break out of the innermost foreach loop
              break
            }
          }
          if { $found_skip == 0 } {
            lappend CCM_info($i,collected_counter_indices)
            $ref_counter
          }
        }
      }
    puts "\nFor CCM $i: list of counters to collect is:\n
    $CCM_info(
    $i,collected_counter_indices)"
      }
    puts "\nTotal perfmon counters for CCM $i,
    $CCM_info($i,CCM_Name): $CCM_info
    ($i,total_counters)\nCounters skipped:
    $CCM_info($i,skipped_counter_indices)"
  }
  set pingCheckStatus 1
  set failedPingCCMList ""
  for { set i 1 } { $i <= $CCM_info(CCM_group_size) }
   { incr i } {
    puts "\nCCM: $i"
    puts "\nChecking ip connectivity on CCM
    $CCM_info($i,CCM_IP)"
    set CCM_info($i,pingCheck) [ pingCheck
     $CCM_info($i,CCM_IP) ]
    if { $CCM_info($i,pingCheck) == 0 } {
      puts "\nPing check failed for CCM
      $CCM_info($i,CCM_IP)"
      incr pingCheckStatus
      lappend failedPingCCMList $i
    } else {
      puts "\nIP connectivity on CCM
```

```
      $CCM_info($i,CCM_IP)
      has been established"
      set CCM_info($i,failedCollections) 0
   }
}
if { $pingCheckStatus == 1 } {
   puts "\nIP connectivity established to all CCMs"
} else {
   puts "\nFailed to establish IP connectivity
   established to all CCMs"
   puts "  - CCMs with that could not be pinged:
   $failedPingCCMList"
   set SOAP_rc 0
}
puts "\nSetting the SOAP debug level and traces"
if { $debug_flag } {
   CCMConUtils::ToggleDebug true
   CCMConUtils::SetDebugInterest $CCMConUtils::
   TraceInterestAll
}
puts "\nSetting the values for CCMConUtils::
PrivateKey and
 CCMConUtils::CertFile"
set CCMConUtils::PrivateKey $CCM_info(key)
set CCMConUtils::CertFile $CCM_info(cert)
if { $SOAP_rc } {
   puts "\nPrepending each perfmon counter with its
   CCM ip address:"
   for { set i 1 } { $i <= $CCM_info(CCM_group_size) }
    { incr i } {
     set CCM_info($i,final_counters) ""
     foreach counter_index $CCM_info
     ($i,collected_counter_indices) {
       set perfmon_counter $perfmon_reference_counters
       ($counter_index)
       if { [ regexp "Cisco MTP Device"
       $perfmon_counter ] &&
            [ info exists CCM_info($i,MTP_Name) ] } {
         regsub -- "MTP_Name" $perfmon_counter
         $CCM_info(
         $i,MTP_Name) perfmon_counter
         puts "Detected a Cisco MTP Device perfmon
         object, substituted its instance name
         with name defined in config file $CCM_info(
         $i,MTP_Name)"
```

```
        }
        if { $CCM_info(nat_flag) } {
          # Use the INSIDE ip address
          set full_counter [ subst -nobackslashes
            {\\$CCM_info(
            $i,CCM_inside_ipaddr)\$perfmon_counter} ]
        } else {
          set full_counter [ subst -nobackslashes
            {\\$CCM_info($i,CCM_IP)\
            $perfmon_counter} ]
        }
    lappend CCM_info($i,final_counters) $full_counter
        if { $debug_flag } {
          puts "Perfmon counter: '$perfmon_counter',
          converted to: '
          $full_counter'"
        }
      }
      if { $debug_flag } {
        # output the complete list of counter for this
          CCM
        puts "\nComplete list of $CCM_info($i,
          total_counters)
        perfmon counters for CCM $i:"
        foreach counter $CCM_info($i,final_counters)
        counter_num
        $CCM_info($i,collected_counter_indices) {
          puts "$counter_num: $counter"
        }
      }
    }
  }
}
if { $SOAP_rc } {
  # flag
  set openSessionStatus 1
  set failedOpenSessionCCMList ""
  for { set i 1 } { $i <= $CCM_info(CCM_group_size) }
  { incr i } {
    puts "\nCalling proc perfmonOpenSession for CCM
      $CCM_info($i,CCM_Name) -
    $CCM_info($i,CCM_IP)"
    set CCM_info($i,perfmonSessionHandle) \
  [ perfmonOpenSession \
                    $CCM_info($i,CCM_Platform) \
                    $CCM_info($i,SOAP_Protocol) \
```

```
                    $CCM_info($i,CCM_IP) \
                    $CCM_info($i,CCM_Username) \
                    $CCM_info($i,CCM_Passwd) \
                    $CCM_info($i,CCM_Name) ]
    if { $CCM_info($i,perfmonSessionHandle) == 0 } {
      puts "\nPerfmon session failed to open for CCM
       $CCM_info($i,CCM_Name) -
      $CCM_info($i,CCM_IP): $CCM_info(
      $i,perfmonSessionHandle)"
      incr openSessionStatus
      lappend failedOpenSessionCCMList $i
    } else {
      puts "\nPerfmon session handle for CCM
       $CCM_info($i,CCM_Name) -
      $CCM_info($i,CCM_IP)
      is: $CCM_info($i,perfmonSessionHandle)"
    }
  }
  if { $openSessionStatus == 1 } {
    puts "\nPerfmon sessions successfully opened
    for all CCMs"
  } else {
    puts "\nFailed to open perfmon sessions for all
      CCMs"
    puts " - CCMs which failed to open a perfmon
    session:
    $failedOpenSessionCCMList"
    set SOAP_rc 0
  }
}
if { $SOAP_rc } {
  # flag
set addCounterStatus 1
  set failedAddCounterCCMList ""
  for { set i 1 } { $i <= $CCM_info(CCM_group_size) }
  { incr i } {
    puts "\nCalling proc perfmonAddCounters for CCM
     $CCM_info(
    $i,CCM_Name) -
    $CCM_info($i,CCM_IP)"
    if { $debug_flag } {
      puts "Adding Perfmon counters to the open
        session
      $CCM_info(
      $i,perfmonSessionHandle):
```

```
          \n$CCM_info($i,final_counters)"
      } else {
        puts "Adding Perfmon counters to the open
        session
        $CCM_info(
        $i,perfmonSessionHandle)"
      }
      if { [ perfmonAddCounters \
                  $CCM_info($i,CCM_Platform) \
                  $CCM_info($i,SOAP_Protocol) \
                  $CCM_info($i,CCM_IP) \
                  $CCM_info($i,CCM_Username) \
                  $CCM_info($i,CCM_Passwd) \
                  $CCM_info($i,CCM_Name) \
                  $CCM_info($i,perfmonSessionHandle) \
                  $debug_flag \
                  $CCM_info($i,final_counters) ] } {
        puts "Successfully added perfmon counters
        on CCM
        $CCM_info(
        $i,CCM_Name) -
        $CCM_info($i,CCM_IP)"
      } else {
        incr addCounterStatus
        lappend failedAddCounterCCMList $i
        puts "Failed to add perfmon counters on CCM
        $CCM_info(
        $i,CCM_Name) -
        $CCM_info($i,CCM_IP)"
        append SOAP_error "Failed to add perfmon
        counters
        on CCM
        $CCM_info($i,CCM_Name) -
        $CCM_info($i,CCM_IP)" "\n"
      }
    }
  if { $addCounterStatus == 1 } {
    puts "\nPerfmon session counters successfully
    added
    for all CCMs"
  } else {
    puts "\nFailed to successfully add perfmon
      session counters
    for all CCMs"
    puts " - CCMs which failed to add perfmon counters:
```

```
        $failedAddCounterCCMList"
        append SOAP_error "Failed to successfully add
        perfmon session counters for all CCMs"
        set SOAP_rc 0
    }
}
if { $SOAP_rc } {
    set temp_result_filename "RESULTS_perfmon_"
    append temp_result_filename "[clock seconds]"
    set temp_result_filename [ file join
    $CCM_info(temp_result_file_path)
    $temp_result_filename ]
    puts "\nFile: $temp_result_filename will contain the
     temporary results,
    to save memory utilization\n"
    set file_created_rc 0
    set file_writing_errors 0
    set file_sourced_rc 1
    if { [ catch { set fh [ open $temp_result_filename w
       ] }
    openerr ] } {
        puts "\nERROR: Failed to open file
         '$temp_result_filename' for
        temporary result storage"
        set file_created_rc 0
    } else {
        puts "\nSuccessfully opened for writing the file '
        $temp_result_filename'
        for temporary result storage"
        # close the file handle for now
        close $fh
        set file_created_rc 1
    }
    # flags
    set collectCounterStatus 1
    set actual_iterations 0
    if { $loop_break == 1 } {
        puts "\n\nInside main for loop, inside if
        loop_break is 1 block:"
        puts "Loop break flag is now 1"
        puts "Current iteration counter i is: $i"
        puts "Current counter actual_iterations is:
        $actual_iterations"
        puts "Current inner for loop iteration counter
            ccm is:
```

```
          $ccm"
          puts "Adjusting counter i value to
            $actual_iterations"
          set i $actual_iterations
          puts "Counter i is now: $i"
          puts "Breaking away from the main for loop"
          break
        }
    set start_clock [clock clicks -milliseconds]
      incr actual_iterations
      puts "\nFor loop iteration: $i out of
      $CCM_info(iterations),
      actual iteration
      $actual_iterations"
      set collectCounterStatus 1
      ############## For loop to collect Perfmon session
      data from EACH CCM node:
      # collect counter values from all the CCMs in the
      CCM group
      for { set ccm 1 } { $ccm <= $CCM_info
        (CCM_group_size) }
       { incr ccm } {
        # reset this iterations formatted result
        set formatted_iteration_result "CCM $CCM_info(
        $ccm,CCM_IP):\n"
        # initialize the new results array element
          indexed by
        (<CCM_number>,<iteration>)
      set CCM_results($ccm,$i) ""
        # call the proc to collect Perfmon session
          counter values
        from CCM
        puts "\nCollecting perfmon counter data for CCM
        $CCM_info($ccm,CCM_IP) -
        $CCM_info($ccm,CCM_Name): $CCM_info(
        $ccm,perfmonSessionHandle)"
        set retval ""
        # record the time we start collecting  began:
        if { ($i == 0) && ($ccm == 1) } {
          set loop_start_time [ clock seconds ]
          puts "\nVery first iteration $i for CCM $ccm,
          $CCM_info($ccm,CCM_Name) -
          $CCM_info($ccm,CCM_IP) started at timestamp:
          $loop_star t_time, [ clock format
          $loop_start_time ]"
```

```
        }
    set soap_transaction_time [clock seconds]
    if { [ expr $CCM_info($ccm,failedCollections)
    > 0 ] } {
      puts "\nCCM $CCM_info($ccm,CCM_Name) had
      previous failed perfmon collections,
      checking if we can ping it before attempting
      a new collection"
      if { ![ pingCheck $CCM_info($ccm,CCM_IP) ] } {
        puts "Skipping this round of perfmon
          collection on CCM
        $CCM_info($ccm,CCM_Name)
        at IP Address $CCM_info($ccm,CCM_IP) since
          it did
        NOT respond to ping"
        set CCM_results($ccm,$i) "ERROR: Failed to
          collect
        perfmon counter values for CCM
        $CCM_info($ccm,CCM_Name) - $CCM_info(
        $ccm,CCM_IP)"
        incr CCM_info($ccm,failedCollections)
puts "Total number of failed perfmon counter
collections for CCM
$CCM_info($ccm,CCM_Name)
is now: $CCM_info($ccm,failedCollections)"
puts "Total number of perfmon session re-open
attempts for CCM
$CCM_info($ccm,CCM_Name)
        is now: $CCM_info($ccm,sessionReOpenAttempts)"
        # skip to the next CCM in this for loop
        continue
      }
    }
if { [ regexp -nocase {[0-9a-f]+-[0-9a-f]+-[0-9a-f]+
-[0-9a-f]+-[0-9a-f]+}
$CCM_info
($ccm,perfmonSessionHandle) ] } {
      if { [ catch { set retval
      [ perfmonCollectSessionCounterData \
            $CCM_info($ccm,perfmonSessionHandle)\
            $CCM_info($ccm,CCM_Platform) \
            $CCM_info($ccm,SOAP_Protocol) \
            $CCM_info($ccm,CCM_IP) \
            $CCM_info($ccm,CCM_Username) \
            $CCM_info($ccm,CCM_Passwd) \
```

```
                    $CCM_info($ccm,CCM_Name) \
                    $debug_flag ]   }
                            perfmon_collect_error ] } {
          puts "ERROR: caught during call to
           perfmonCollectSessionCounterData:\n
          $perfmon_collect_error"
           set retval ""
       } else {
         if { [ string match $retval "1" ] } {
         set CCM_info($ccm,failedCollections) 0
         set CCM_info($ccm,sessionReOpenAttempts) 0
         }
       }
     } else {
       puts "\nSkipping perfmon collection since we
       do NOT have a
       valid perfmon
       session handle yet: '$CCM_info(
       $ccm,perfmonSessionHandle)'"
       set retval ""
     }
     if { $i == 0 } {
 puts "\nFirst iteration $i for CCM $ccm,
 $CCM_info($ccm,CCM_Name) -
 $CCM_info($ccm,CCM_IP):
 tossing out the very first set of perfmon values"
       if { $debug_flag } {
         puts "\nVery First set of perfmon values:
         \n$retval"
       }
       puts "\nImmediately resubmitting the
       transaction and saving the
        next set of values"
       set soap_transaction_time [clock seconds]
       set retval [ perfmonCollectSessionCounterData\
           $CCM_info($ccm,perfmonSessionHandle) \
           $CCM_info($ccm,CCM_Platform) \
           $CCM_info($ccm,SOAP_Protocol) \
           $CCM_info($ccm,CCM_IP) \
           $CCM_info($ccm,CCM_Username) \
           $CCM_info($ccm,CCM_Passwd) \
           $CCM_info($ccm,CCM_Name) \
           $debug_flag ]
       if { $debug_flag } {
         puts "\nRecall of proc
```

```
                    perfmonCollectSessionCounterData
                    returned:\n
                    $retval"
                }
            }
        # end of if $i == 0
        # Only if we got meaningful results from
        perfmon collection, split list on commas
        if { ![ string match $retval "" ] &&
        $retval != 0 } {
            # debug
            if { $debug_flag } {
                puts "\nLength of returned list of collected
                perfmon counters:
                [ llength $retval ]"
            }
            set CCM_results($ccm,$i) $retval
foreach counter_result $retval {
                if { $debug_flag } {
                    puts "Processing Counter result
                    $counter_result"
                }
                # split on commas
                set split_counter_result [ split
                $counter_result "," ]
                set full_counter_name [ lindex
                $split_counter_result 0 ]
                if { $debug_flag } {
                    puts "Full counter name:
                    $full_counter_name"
                }
                # pull the counter name out
                set just_counter ""
                if { [ regexp {\\\\[0-9.]+\\Cisco MTP Device
                \(.*\)\\(.*)$}
                $full_counter_name trash mtp_counter ]
                } {
                    set just_counter [ subst -nobackslashes
                    {Cisco MTP Device(MTP_Name)\
                    $mtp_counter} ]
                    if { $debug_flag } {
                puts "Detected a Cisco MTP Device counter:
                $full_counter_name."
                        puts "Modified counter name for sorting
                        purposes to:
```

```
                    $just_counter"
                }
            } else {
                regexp {\\\\[0-9.]+\\(.*)}
                $full_counter_name trash
                just_counter
            }
            if { $debug_flag } {
                puts "Counter name: $just_counter"
            }
            set counter_cstatus [ lindex
            $split_counter_result 2 ]
            set counter_value [ lindex
            $split_counter_result 1 ]
            if { $counter_cstatus == 0 ||
            $counter_cstatus == 1 } {
                if { $debug_flag } {
                    puts "Counter value: $counter_value"
                }
            } else {
                puts "ERROR: counter '$just_counter'
                CStatus value is NOT 0 or 1:
                $counter_cstatus"
                puts "Invalid perfmon counter value
                collected:
                $counter_value"
                append counter_value "-" "ERROR"
                puts "Counter value now set to:
                    $counter_value"
            }
            if { $debug_flag } {
                puts "Counter cstatus: $counter_cstatus"
            }
    foreach index $CCM_info(
    $ccm,collected_counter_indices) {
            # since we have back slashes,
            string equal must be used
            # instead of string match
            if { [ string equal
            $perfmon_reference_counters(
            $index) $just_counter ] } {
                if { $debug_flag } {
                    puts "Matched perfmon counter '
                    $perfmon_reference_counters($index)'
                    at index: $index \n"
```

```
                      }
                      set CCM_perfmon_results($index)
                      $counter_value
                    }
                  }
                }
                set ordered_counter_values {}
                foreach index $CCM_info(
                $ccm,collected_counter_indices) {
                  lappend ordered_counter_values
                  $CCM_perfmon_results($index)
                }
                if { $debug_flag } {
                  puts "Ordered list of counter values: \n
                  $ordered_counter_values"
                }
      set count 0
                foreach counter_index $CCM_info(
                $ccm,collected_counter_indices) {
                set temp_results($counter_index) [ lindex
                $ordered_counter_values $count ]
                  if { $debug_flag } {
                    puts "Temp results array temp_results
                    ($counter_index) =
                    $temp_results($counter_index)"
                  }
                  incr count
                }
                foreach counter_index $CCM_info(
                $ccm,skipped_counter_indices) {
                  set temp_results($counter_index) "NA"
                  if { $debug_flag } {
                    puts "Temp results array temp_results
                    ($counter_index) =
                    $temp_results($counter_index)"
                  }
                }
          }
                set corrected_retval [ join $corrected_retval
                  "," ]
                set CCM_results($ccm,$i) $corrected_retval
                append CCM_results($ccm,$i) ","
                $soap_transaction_time
                set formatted_soap_transaction_time
                  [ clock format
```

```
      $soap_transaction_time ]
      puts "$formatted_soap_transaction_time:
        Iteration
      $i out of $CCM_info(iterations),
      CCM $ccm, $CCM_info($ccm,CCM_Name) -
      $CCM_info(
      $ccm,CCM_IP):\n$CCM_results($ccm,$i)"
      set CCM_info($ccm,failedCollections) 0
      set CCM_info($ccm,sessionReOpenAttempts) 0
      if { $debug_flag } {
        for { set k 0 } { $k < $CCM_info
        ($ccm,total_counters) }
        { incr k } {
          append formatted_iteration_result "\n"
          "Counter: [ lindex
          $CCM_info
          ($ccm,final_counters) $k ],"
          append formatted_iteration_result "
          Value: [ lindex
          $ordered_counter_values $k]"
}

        # output theformatted iteration result for
          this CCM
        puts "\n$formatted_iteration_result"
      }
    # if retval from call to proc
    perfmonCollectSessionCounterData failed
    } else {
      puts "\nFailed to collect perfmon counter
      values for CCM
      $CCM_info($ccm,CCM_Name) -
      $CCM_info($ccm,CCM_IP):"
      puts "Proc perfmonCollectSessionCounterData
       returned:\n
      $retval"
      set collectCounterStatus 0
      set CCM_results($ccm,$i) "ERROR: Failed to
      collect
      perfmon counter values for CCM
      $CCM_info($ccm,CCM_Name) -
      $CCM_info($ccm,CCM_IP)"
      incr CCM_info($ccm,failedCollections)
      puts "Total number of failed perfmon counter
      collections for CCM
      $CCM_info($ccm,CCM_Name) is now:
```

```
            $CCM_info(
            $ccm,failedCollections)"
            puts "Total number of perfmon session re-open
              attempts for CCM
            $CCM_info($ccm,CCM_Name) is now: $CCM_info(
            $ccm,sessionReOpenAttempts)"
if { ([ expr $CCM_info
($ccm,failedCollections) >=
$min_failed_collections ]) &&
                ([ pingCheck $CCM_info($ccm,CCM_IP) ]) &&
                ([ expr $CCM_info
                ($ccm,sessionReOpenAttempts) <=
                $max_reopen_session_attempts ]) } {
            puts "We have exceeded $min_failed_
            collections failed perfmon counter
            collections for CCM$CCM_info($ccm,CCM_Name),
            attempting
to close current perfmon session and re-open a
new one and add back the counters"
            puts "Calling proc perfmonCloseSession for
            CCM
            $ccm,
            $CCM_info($ccm,CCM_Name)
            - $CCM_info($ccm,CCM_IP)"
            if { ![ string match $CCM_info(
            $ccm,perfmonSessionHandle) "0" ] &&
                ![ string match $CCM_info(
                $ccm,perfmonSessionHandle) "" ] } {
              catch { perfmonCloseSession \
                    $CCM_info($ccm,CCM_Platform) \
                    $CCM_info($ccm,SOAP_Protocol) \
                    $CCM_info($ccm,CCM_IP) \
                    $CCM_info($ccm,CCM_Username) \
                    $CCM_info($ccm,CCM_Passwd) \
                    $CCM_info($ccm,CCM_Name) \
                    $CCM_info($ccm,perfmonSessionHandle) }
            }
            puts "Calling proc perfmonOpenSession for
            CCM
            $CCM_info($ccm,CCM_Name) -
            $CCM_info($ccm,CCM_IP)"
            # need to have catch here otherwise script
              will crash if CM server not responding
            set CCM_info($ccm,perfmonSessionHandle) 0
            catch { set CCM_info($ccm,
```

```
                        perfmonSessionHandle) \
                        [ perfmonOpenSession \
                          $CCM_info($ccm,CCM_Platform) \
                          $CCM_info($ccm,SOAP_Protocol) \
                          $CCM_info($ccm,CCM_IP) \
                          $CCM_info($ccm,CCM_Username) \
                          $CCM_info($ccm,CCM_Passwd) \
                          $CCM_info($ccm,CCM_Name) ] }
        if { [ string match $CCM_info($ccm,
            perfmonSessionHandle) "0" ] ||
                    [ string match $CCM_info(
                    $ccm,perfmonSessionHandle) "" ]  } {
                puts "Perfmon session failed to re-open
                for CCM
                $CCM_info(
                $ccm,CCM_Name) -
                $CCM_info($ccm,CCM_IP): $CCM_info
                ($ccm,perfmonSessionHandle)"
                incr CCM_info($ccm,sessionReOpenAttempts)
              } else {
                puts "New perfmon session handle for CCM
                $CCM_info(
                $ccm,CCM_Name) -
                $CCM_info($ccm,CCM_IP) is: $CCM_info
                ($ccm,perfmonSessionHandle)"
                # now add back the perfmon counters to
                  this newly
                opened perfmon session
                puts "Attempting to add back the perfmon
                  counters
                  to the newly
                opened session"
                if { [ perfmonAddCounters \
                        $CCM_info($ccm,CCM_Platform) \
                        $CCM_info($ccm,SOAP_Protocol) \
                        $CCM_info($ccm,CCM_IP) \
                        $CCM_info($ccm,CCM_Username) \
                        $CCM_info($ccm,CCM_Passwd) \
                        $CCM_info($ccm,CCM_Name) \
                        $CCM_info($ccm,perfmonSessionHandle) \
                        $debug_flag \
                        $CCM_info($ccm,final_counters) ] } {
                    puts "Successfully added back perfmon
                      counters on CCM
                    $CCM_info($ccm,CCM_Name) -
```

```
                    $CCM_info($ccm,CCM_IP)"
                    # reset the counters
                    set CCM_info($ccm,failedCollections) 0
                    set CCM_info($ccm,sessionReOpen
                    Attempts)0
                  } else {
                    puts "Failed to add perfmon counters
                    on CCM
                    $CCM_info($ccm,CCM_Name) -
                    $CCM_info($ccm,CCM_IP)"
                  }
}
          # end of else of if perfmonSessionHandle==0
        }
      }
      # end of else of if retval from call to proc
      perfmonCollectSessionCounterData failed
    }
    # end of inner for loop processing each CCM in the
    CCM group
    if { ($file_created_rc == 1) && ([ expr fmod(
    $i,10) == 0.0 ]) } {
      puts "\n############# Iteration: $i,
      Writing contents of array CCM_results()
      to file $temp_result_filename"
      set ts [ clock seconds ]
      set formatted_ts [ clock format $ts ]
      set write_result 0
      set write_result [ catch {
        # open file for appending to it
        set fh [ open $temp_result_filename a ]
        set CCM_results_array_elems [ array names
          CCM_results ]
        puts $fh "\n## Iteration: $i\n"
foreach elem $CCM_results_array_elems {
          ## puts $fh "set CCM_written_results($elem)
          $CCM_results($elem)"
          puts $fh "set CCM_written_results($elem) \"
          $CCM_results($elem)\""
        }
        # flush the write buffer
        flush $fh
        # close the file handle
        close $fh
      } write_result_err ]
```

```
      #check if we hit any error while writing results
      to a file
      if { $write_result ne "0"  } {
        puts "\nERROR encountered writing results
          to file
        $temp_result_filename,
        at iteration $i :"
        puts "$write_result_err\n"
        incr file_writing_errors
      } else {
        puts "\n###### Successfully completed writing
        contents of array
        CCM_results() to file $temp_result_filename"
        puts "##### Clearing array CCM_results()
          contents,
        last iteration: $i \n"
        # UNSET the array and free up its memory
        catch { array unset CCM_results }
      }
    }
  set end_clock [clock clicks -milliseconds]
    set sleep_delay [expr ($CCM_info(polling_interval) *
    1000) -
    abs($end_clock - $start_clock)]
    if {$sleep_delay > 0} {
        puts "\nSleeping for
        $CCM_info(polling_interval)
        seconds (actual:
        $sleep_delay ms)"
        after $sleep_delay
    } else {
        puts "\nWARNING: Counter gathering took
        longer than the poll delay! -
        Sleeping 1 second anyway"
        after 1000
    }
    ## puts "\nSleeping for
    $CCM_info(polling_interval) seconds"
    ## sleep $CCM_info(polling_interval)
  }
  # end of Main outer for loop
\
  puts "\nExited from the main for loop at
  timestamp: [clock seconds]"
  puts "Main for loop counter i is: $i,
```

```
    counter actual_iterations is:
    $actual_iterations"
    puts "Inner for loop counter ccm is: $ccm"
    puts "Total iterations scheduled:
    $CCM_info(iterations),
    total iterations processed:
    $actual_iterations\n"
} else {
  # end of if SOAP_rc
  puts "\nERROR: No perfmon counter values collected"
  set execution_status "ERRORED"
}
puts "\n###Overall Perfmon Counter Statistics ##"
# return the final list of counter values
set overall_result "\nFINAL RESULTS:\n"
append overall_result "===============" "\n"
# Build the FINAL RESULTS section ONLY if we
were successful at collecting
# perfmon counter data via SOAP
if { $SOAP_rc == 1 } {
  append overall_result "\nPerfmon Legend Begin" "\n"
  foreach ref_counter $reference_counters_indices {
    append overall_result $ref_counter ","
    $perfmon_reference_counters
    ($ref_counter) "\n"
  }
  append overall_result "Perfmon Legend End\n\n"
  append overall_result "\n# CallManager name,
  iteration,
  perfmon counter 0,perfmon
  counter 1,....,perfmon counter n,timestamp"
  append overall_result "\nPerfmon Data Begin" "\n"
  if { [ info exists CCM_results ] } {
      set write_result 0
      set write_result [ catch {
        # open file for appending to it
        set fh [ open $temp_result_filename a ]
        set CCM_results_array_elems [ array names
        CCM_results ]
        puts $fh "\n## Adding contents of what is
        left in CCM_result array.
        Last iteration was: $i\n"
        foreach elem $CCM_results_array_elems {
          puts $fh "set CCM_written_results($elem) \"
          $CCM_results($elem)\""
```

```
            }
            # flush the write buffer
            flush $fh
            # close the file handle
            close $fh
        } write_result_err ]
        # check if we hit any error while writing
        results to a file
        if { $write_result ne "0"  } {
          puts "\nERROR encountered writing LAST
          batch of results to file
          $temp_result_filename,
          last iteration was: $i"
  puts "$write_result_err\n"
          # continue to store the results in array
          CCM_results()
           locally for
          the rest of the script
          incr file_writing_errors
        }
    }
    puts "\nSleeping 15 seconds before sourcing the
     temp result file:
    '$temp_result_filename'"
    sleep 15
    if { [ catch { source $temp_result_filename }
    source_err ] } {
      puts "\nERROR: failed to source
      temp_result_filename '
      $temp_result_filename':\n$source_err"
      set file_sourced_rc 0
    } else {
      puts "\nSuccessfully sourced
      temp_result_filename '
      $temp_result_filename'"
      set file_sourced_rc 1
    }
} # end of if ($file_created_rc == 1)

for { set ccm 1 } { $ccm <= $CCM_info
(CCM_group_size) }
{ incr ccm } {
  for { set i 0 } { $i < $actual_iterations }
  { incr i } {
    set result_line $CCM_info($ccm,CCM_Name)
```

```
      if { $file_created_rc == 0 } {
        append result_line "," $i "," $CCM_results(
        $ccm,$i)
        } elseif { $file_sourced_rc == 1 } {
        append result_line "," $i ","
         $CCM_written_results($ccm,$i)
        }
        append overall_result $result_line "\n"
  }
  }
  append overall_result "Perfmon Data End\n\n"
## end of if { $SOAP_rc == 1 }
} else {
  # if SOAP_rc is NOT 1:
  append overall_result "\nERROR: no perfmon
  data collected" "\n"
  $SOAP_error
  set execution_status "FAIL"
}
if { $file_writing_errors != 0 } {
  append overall_result "\nERROR: perfmon counter
  data were UN-RELIABLY written "
  append overall_result "to a temp file and should
  not be trusted.
  Encountered "
  append overall_result "$file_writing_errors file
  writing errors! \n"
  set execution_status "FAIL"
}
append overall_result "\n"
 "END OF FINAL RESULTS" "\n"
# output the overall result
puts $overall_result
#######Close Perfmon Session #######
if { $SOAP_rc } {
  # flag
  set closeSessionStatus 1
  set failedCloseSessionCCMList ""
for { set i 1 } { $i <= $CCM_info(CCM_group_size) }
 { incr i } {
    # call the proc to close a Perfmon session
     to CCM
    puts "\nCalling proc perfmonCloseSession
    for CCM $i,
    $CCM_info($i,CCM_Name) -
```

```
    $CCM_info($i,CCM_IP)"
    if { [ perfmonCloseSession \
                    $CCM_info($i,CCM_Platform) \
                    $CCM_info($i,SOAP_Protocol) \
                    $CCM_info($i,CCM_IP) \
                    $CCM_info($i,CCM_Username) \
                    $CCM_info($i,CCM_Passwd) \
                    $CCM_info($i,CCM_Name) \
                    $CCM_info($i,perfmonSessionHandle)
                    ] } {
        puts "Perfmon session for CCM $i, $CCM_info(
        $i,CCM_Name) -
        $CCM_info($i,CCM_IP)
        $CCM_info($i,perfmonSessionHandle)
        successfully closed"
    } else {
        incr closeSessionStatus
        lappend failedCloseSessionCCMList $i
        puts "Perfmon session failed toclosefor CCM $i,
        $CCM_info(
        $i,CCM_Name) - CCM
        $CCM_info($i,CCM_IP): $CCM_info($i,
        perfmonSessionHandle)"
    }
  }
  if { $closeSessionStatus == 1 } {
    puts "\nPerfmon sessions successfully closed for
    all CCMs"
  } else {
    puts "\nFailed to successfully close perfmon
    sessions for all CCMs"
    puts " - CCMs which failed to close
    perfmon session:
    $failedCloseSessionCCMList"
 set SOAP_rc 0
  }
}

if { [ string match $execution_status "FAIL" ] } {
  puts "\nSCRIPT EXECUTION STATUS: FAIL"
} else {
  puts "\nSCRIPT EXECUTION STATUS: PASS"
}

# record the time the script ended:
```

```
set stop_time [ clock seconds ]
puts "\n\nScript [ info script ] $argv is done
at timestamp:
$stop_time,
[ clock format $stop_time ]"
```

### 13.1.3   SIP Phone Registration Source Code

```
#!/bin/sh
# \
exec $AUTOTEST/bin/expect "$0" "$@"
################################
#
# File: SIP_Registration.tcl
#
# Author: Sureshkumar V. Subramanian
################################
set start_time [ clock seconds ]
puts "Script [ info script ] $argv started at
timestamp:
$start_time,
[ clock format $start_time]"
puts "ARGV is: '$argv', ARGC is: '$argc'"
puts "lindex argv 0 is: [ lindex $argv 0 ]"
if { $argc < 1 } {
  puts "ERROR: expecting at least one command
  line argument, the config file to source"
  puts "ARGC: $argc"
  puts "ARGV: $argv"
  puts "Aborting the script"
  return
}
set cfg_file [ lindex $argv 0 ]
puts "Sourcing the config file passed in the first
command line argument: '$cfg_file'"
if { [ catch { source $cfg_file } errmsg ] } {
  puts "ERROR: in sourcing the config file
   '$cfg_file': $errmsg"
 puts "Aborting the script"
  return
}
puts "\nSuccessfully sourced the config file
 '$cfg_file'"
```

```tcl
if { [ info exists camelot_cfg(lib_file) ] } {
  puts "Sourcing the library file
  '$camelot_cfg(lib_file)' as defined
  in the config file"
  if { [ catch { source $camelot_cfg(lib_file) }
   errmsg ] } {
    puts "ERROR: in sourcing the
    camelot_cfg(lib_file) '
    $camelot_cfg(lib_file)': $errmsg"
    puts "Aborting the script"
    return
  } else {
    puts "\nSuccessfully sourced the library file '
    $camelot_cfg(lib_file)'"
  }
} else {
  puts "ERROR: Required variable
  'camelot_cfg(lib_file)'
  is NOT defined in the config file"
  puts "Aborting the script"
  return
}
set main_loop_break 0
set env(TERM) "vt100"
set expect_library [ file join
$env(AUTOTEST) local lib ]
source [ file join $expect_library control.exp ]
log_user 0
set camelotvapiei_pkg_ver [package
 require camelotvapiei]
puts "[getTime] :INFO: package camelotvapiei
version
 $camelotvapiei_pkg_ver has been required"
set bcg_pkg_ver [package require bcg ]
puts "[getTime] :INFO: package bcg version
$bcg_pkg_ver has been required"
set script_pid [ pid ]
puts "[getTime] :INFO: Process ID of
this script is: $script_pid"
if { 0 } {
set start_camserv 1
set camelot_cfg(cam_id) "SCAL-CAM2"
set camelot_cfg(cam_ip) "1.1.1.103"
set camelot_cfg(cam_port) "7000"
set camelot_cfg(cam_username) "lab"
```

```
set camelot_cfg(cam_passwd) "lab"
set reg_group_size 10
set deviceid 1
puts "[getTime] :INFO: Installing trap to
break out of main program loop"
trap {
  set trap_name [trap -name]
  set trap_ts [ clock seconds]
  set trap_fts [ clock format $trap_ts]
  puts "[getTime] :WARNING: got signal named '
  $trap_name' at $trap_fts"
   set main_loop_break 1
} {HUP INT QUIT PIPE TERM CONT WINCH
USR1 USR2}
###############################################
## Start Camelot Server on one port on the
Camelot PC
###############################################
set camserv_started ""
if { $start_camserv == 1 } {
  puts "[getTime] :INFO: Telnetting to ip addr
  $camelot_cfg(cam_ip) to start camserv
  on Camelot PC $camelot_cfg(cam_id) on port
  $camelot
_cfg(cam_port)"
  set camserv_started ""
  if { [ info exists camelot_cfg(sd_option) ] &&
      [ string match $camelot_cfg(sd_option) "1"]}{
  puts "[getTime] :INFO: starting camserv with -
  sd option in order to capture SIP
  Messages on camserv, and have them displayed later
  in VLOG traces"
set camserv_started [ start_camelot_server
$camelot_cfg(cam_id) \
                        $camelot_cfg(cam_ip) \
                        $camelot_cfg(cam_username) \
                        $camelot_cfg(cam_passwd) \
                        $camelot_cfg(cam_port) \
                        $camelot_cfg(sd_option) ]
  # start camserv without the -sd option:
  } else {
    set camserv_started [ start_camelot_server
    $camelot_cfg(cam_id) \
                        $camelot_cfg(cam_ip) \
                        $camelot_cfg(cam
```

```
                                              _username)\
                              $camelot_cfg(cam_passwd)\
                              $camelot_cfg(cam_port) ]
  }
  # verify that camserv started successfully
  if { $camserv_started == 1 } {
    puts "[getTime] :INFO: Camelot server started
     successfully"

  } else {
    append error_message "\n[getTime] :ERROR:
    Camelot server failed to start
    successfully on:"
    append error_message "\n - PC:
    $camelot_cfg(cam_pc)"
    append error_message "\n - ip:
    $camelot_cfg(cam_ip)"
    append error_message "\n - port:
    $camelot_cfg(cam_port)"
    append error_message "\n - username:
    $camelot_cfg(cam_username)"
    append error_message "\n - passwd:
    $camelot_cfg(cam_passwd)"

    append error_message "\nAborting the script"
    puts $error_message
    exit
  }

}
set init_proc_call(sip,0) {init_legacysip_endpoints
 $camelot_cfg(cam_ip)
$camelot_cfg(cam_port) $ccm_ip $mac_addr_prefix
 $mac_addr_start $
dn_start $ep_index_begin $ep_index_end
$deviceid $tftpip $sip_port}
## 08/01/07:
# Flag to signal if we should abort due to a
configuration error
set abort_flag 0

## 05/12/06:
set ep_index_begin_list {}
set ep_index_end_list {}
set total_endpoints 0
```

```
foreach cfg_batch $init_batch_list {

  puts "\nINFO: Initializing endpoints in scenario
   defined by:
   $cfg_batch"

  # split the list on commas
  set split_cfg_batch [ split $cfg_batch "," ]
  set ccm_ip [ lindex $split_cfg_batch 0 ]
  set mac_addr_prefix [ lindex $split_cfg_batch 1 ]
  set mac_addr_start [ lindex $split_cfg_batch 2 ]
  set dn_start [ lindex $split_cfg_batch 3 ]
  set total_lines [ lindex $split_cfg_batch 4 ]
  set line_dn_increment [ lindex $split_cfg_batch 5 ]
  set ep_index_begin [ lindex $split_cfg_batch 6 ]
  set ep_index_end [ lindex $split_cfg_batch 7 ]
  lappend ep_index_begin_list $ep_index_begin
  lappend ep_index_end_list $ep_index_end
  set ep_type [ lindex $split_cfg_batch 8 ]
  set ep_transport [ lindex $split_cfg_batch 9 ]
  set deviceid [ lindex $split_cfg_batch 10 ]
  set security_mode [ lindex $split_cfg_batch 11 ]
  set gencert_flag [ lindex $split_cfg_batch 12 ]
if { $ep_index_begin == $ep_index_end } {
    set total_scenario_eps 1

  } elseif { [ expr $ep_index_end >
  $ep_index_begin ] } {
    set ep_diff [ expr $ep_index_end -
    $ep_index_begin ]
    set total_scenario_eps [ expr
    $ep_diff + 1 ]

  } else {
    puts "ERROR: the ep_index_end MUST be >= of
    ep_index_begin"
    puts "        We have ep_index_end
    $ep_index_end which is SMALLER than
    ep_index_begin $ep_index_begin"
    puts "Aborting script"
    set abort_flag 1
    # break out of the foreach loop
    break
  }
```

```
  puts "INFO: Checking if scenario endpoint Total
  $total_scenario_eps
  is divisible the reg_group_size $reg_group_size"
   if { [ expr fmod($total_scenario_eps,
   $reg_group_size) == 0.0 ] } {
     puts "INFO: Validated that Scenario
     endpoint Total
     $total_scenario_eps
     is divisible the reg_group_size
     $reg_group_size"
   } else {
     puts "ERROR: Scenario endpoint Total
     $total_scenario_eps is NOT divisible
     the reg_group_size $reg_group_size !"
     puts "Aborting script"
     set abort_flag 1
     # break out of the foreach loop
     break
   }
 puts "[getTime] :INFO: Initializiation of
 $total_scenario_eps Endpoints Begins"
   set total_endpoints [ expr $total_endpoints +
   $total_scenario_eps ]
   puts "[getTime] :INFO: Total endpoints
   calculated so far is:
   $total_endpoints"
   }
   } else {
     puts "[getTime] :INFO: Initializiation of
     $ep_type Endpoints
     $ep_index_begin to
     $ep_index_end completed"
   }
}
if { [ string match $abort_flag "1" ] } {
  puts "[getTime] :WARNING: Stopping Camelot
  server before Aborting"
  if { $camserv_started == 1 } {
  set camserv_stopped [ stop_camelot_server
  $camelot_cfg(cam_id) ]
     if { $camserv_stopped == 1 } {
       puts "[getTime] :INFO: Camelot server
       $camelot_cfg(cam_id) stopped
       successfully"
     } else {
```

```
        puts "[getTime] :ERROR: Camelot server
        $camelot_cfg(cam_id)
        did not stop gracefully,
        exiting script anyway!"
      }
   }
   puts "[getTime] :INFO: Script [ info script ]
   Aborted"
   # exit the script
   return
}
puts "[getTime] :INFO: Bringing $total_endpoints
 endpoints in service"
foreach scen_ep_begin
$ep_index_begin_list scen_ep_end
$ep_index_end_list {
   puts "[getTime] :INFO: Calling alter_endpoints
   $scen_ep_begin $scen_ep_end
   inservice $reg_group_size"
   alter_endpoints $scen_ep_begin
   $scen_ep_end "inservice" $reg_group_size
}
puts "[getTime] :INFO: Registration of
$total_endpoints
Endpoints Finished"
puts "[getTime] :INFO: Begining main
control loop f
or process id $script_pid"
set while_iteration 0
while { $main_loop_break == 0 } {
   set main_ts [ clock seconds ]
   set main_fts [ clock format $main_ts ]
   incr while_iteration
   puts "[getTime] :INFO: In while loop,
   iteration
   $while_iteration"
   puts "[getTime] :INFO: Checking to see if all
   endpoints are inservice"
   foreach scen_ep_begin
   $ep_index_begin_list scen_ep_end
   $ep_index_end_list {
   array set ep_status [ check_endpoints_state
   $scen_ep_begin
   $scen_ep_end "inservice" ]
      if { $ep_status(status) == 1} {
```

```
        puts "[getTime] :INFO: All endpoints
         ranging from index
        $scen_ep_begin to
        $scen_ep_end are inservice"
      } else {
        puts "[getTime] :INFO: endpoints not inservice:
        $ep_status(not_in_desired_state)"
      }
    }
    set minutes [ expr $camelot_cfg(iteration_hold) / 60 ]
    puts "[getTime] :INFO: Sleeping for
    $minutes minutes..."
    sleep $camelot_cfg(iteration_hold)
}

puts "[getTime] :INFO: Broke out of main loop
 at iteration
 $while_iteration"
puts "[getTime] :INFO: Bringing
$total_endpoints endpoints out-of-service"
foreach scen_ep_begin
$ep_index_begin_list scen_ep_end
$ep_index_end_list {
  puts "[getTime] :INFO: Calling alter_endpoints
  $scen_ep_begin
  $scen_ep_end
  outofservice $reg_group_size"
  alter_endpoints $scen_ep_begin
  $scen_ep_end "outofservice"
  $reg_group_size
}
foreach scen_ep_begin
$ep_index_begin_list scen_ep_end
$ep_index_end_list {
  puts "[getTime] :INFO: Calling uninit_endpoints
  $scen_ep_begin $scen_ep_end"
  uninit_endpoints $scen_ep_begin
  $scen_ep_end
}
foreach scen_ep_begin
$ep_index_begin_list scen_ep_end
$ep_index_end_list {
  array set final_ep_state [ check_endpoints_state
  $scen_ep_begin
  $scen_ep_end "uninitialized" ]
```

```
    if { $final_ep_state(status) == 1 } {
      puts "[getTime] :INFO: Endpoints
      $scen_ep_begin to
      $scen_ep_end are now uninitialized"
    } else {
      puts "[getTime] :WARNING: The following endpoints
      were NOT successfully
      uninitialized:\n$final_ep_state
      (not_in_desired_state)"
    }
  }
  puts "[getTime] Attempted to uninitialize all
  $total_endpoints endpoints.
  Attempting to stop camserv.exe"
  set camserv_stopped [ stop_camelot_server
  $camelot_cfg(cam_id) ]

    if { $camserv_stopped == 1 } {
      puts "[getTime] :INFO: Camelot server
      $camelot_cfg(cam_id)
      stopped successfully"
    } else {
      puts "[getTime] :INFO: Camelot server
      $camelot_cfg(cam_id)
      did not stop gracefully,
      exiting script anyway!"
    }
  }
  puts "[getTime] :INFO: Script [ info script ]
  stopped"
```

### 13.1.4  *Place Call Source Code*

```
#!/bin/sh
# \
exec $AUTOTEST/bin/expect "$0" "$@"
#####################################
#
# File: Place_Call.tcl
#
# Author: Sureshkumar V. Subramanian
#####################################
```

```tcl
set camelot_cfg(lib_file) "/opt/modela/
Suresh/scalnewfiles/
vapi-ei/modelaVapieiLIB_suresh.exp"
source $camelot_cfg(lib_file)
set rc 1
set error_messages ""
set start_time [ clock seconds ]
puts "[getTime] :INFO: Script [ info script ] started
at timestamp: $start_time,
[ clock format $start_time ]"
puts "[getTime] :INFO: Installing trap to break out
of main program loop"
trap {
  set trap_name [trap -name]
  set trap_ts [ clock seconds]
  set trap_fts [ clock format $trap_ts]
  puts "[getTime] :WARNING: got signal named '
  $trap_name' at $trap_fts"
 set main_loop_break 1
} {HUP INT QUIT PIPE CONT WINCH USR1 USR2}
set main_loop_break 0
log_user 0
set cam_pkg [package require camelotvapiei]
puts "[getTime] :INFO: package camelotvapiei
$cam_pkg has been required"
set bcg_pkg [package require bcg]
puts "[getTime] :INFO: package bcg
$bcg_pkg has been required"
set script_pid [ pid ]
puts "[getTime] :INFO: Process ID
of this script is: $script_pid"
set required_argv_len 7
if { [ expr $argv_len < $required_argv_len ] } {
  append error_messages "\nERROR:
   incorrect number
  of arguments passwd to this script!"
  append error_messages "\n
  Expecting at least
  $required_argv_len arguments,
  and got the following $argv_len arguments:"
  append error_messages "\n"
  set cmd_arg_counter 0
  foreach cmd_arg $argv {
    append error_messages "\n - Argv index
    $cmd_arg_counter:
```

```
        $cmd_arg"
        incr cmd_arg_counter
    }
    append error_messages "\n\nAborting script."
    puts $error_messages
    return
} else {
    # check if autofeature_flag is set to 1
    if { [ lindex $argv 8 ] == 1 } {
        set required_argv_len 8
        if { [ string match [ lindex $argv 7 ] "" ] } {
            append error_messages "\nERROR:
            incorrect number
            of arguments passwd to this script!"
            ## append error_messages "\n Expecting a
            10th argument to be passed,
            which contains the autofeatures config
            parms, and got:
            [lindex $argv 9 ]"
            append error_messages "\nExpecting
            an 8th argument
            to be passed, which contains the
            autofeatures config parms, and got:
              [ lindex $argv 7 ]"
            append error_messages "\nAborting script."
            puts $error_messages
            return
        }
    }
}
set camserv_parms [split [lindex $argv 0] ":" ]
set camelot_cfg(cam_orig_ip)   [lindex
$camserv_parms 0]
set camelot_cfg(cam_orig_port) [lindex
$camserv_parms 1]
set camelot_cfg(cam_term_ip)   [lindex
$camserv_parms 2]
set camelot_cfg(cam_term_port) [lindex
$camserv_parms 3]
set bcg_parms [split [lindex $argv 1] ":" ]
set camelot_cfg(bcg_ip)   [lindex
$bcg_parms 0]
set camelot_cfg(bcg_port) [lindex
  $bcg_parms 1]
# BCG Call rate in Calls Per Seconds
```

```
set camelot_cfg(bcg_call_rate) [lindex
  $bcg_parms 2]
# Call Hold Time (in miliseconds) 1000 = 1 second
set camelot_cfg(call_holdtime) [lindex
$bcg_parms 3]
set ep_orig_term_ranges
[split [lindex $argv 2] ":" ]
set camelot_cfg(ep_orig_start)      [lindex
  $ep_orig_term_ranges 0]
set camelot_cfg(ep_orig_end)        [lindex
$ep_orig_term_ranges 1]
set camelot_cfg(orig_line_number)   [lindex
$ep_orig_term_ranges 2]
set camelot_cfg(ep_term_start)      [lindex $
ep_orig_term_ranges 3]
set camelot_cfg(ep_term_end)        [lindex
$ep_orig_term_ranges 4]
set camelot_cfg(term_line_number)   [lindex
 $ep_orig_term_ranges 5]
set camelot_cfg(base_dn)            [lindex
 $ep_orig_term_ranges 6]
set camelot_cfg(autoanswer_delay)   [lindex
$ep_orig_term_ranges 7]
set camelot_cfg(shared_line_group_size) [lindex
$ep_orig_term_ranges 8]
} else {
  set camelot_cfg(shared_line_flag) 0
  set camelot_cfg(shared_line_group_size) 1
  puts "[getTime] :INFO: Setting Shared Line
   flag to 0,
  since no shared line
  group size value was specified"
set iteration_parms [split [lindex $argv 3] ":" ]
set camelot_cfg(total_iterations)    [lindex
$iteration_parms 0]
set camelot_cfg(iteration_interval) [lindex
$iteration_parms 1]
set camelot_cfg(test_name) [lindex $argv 4]
set auto_path_conf_list [split [lindex $argv 5] ":" ]
set camelot_cfg(pathconf_flag) [lindex
 $auto_path_conf_list 0 ]
set camelot_cfg(pathconf_ep_skip_factor)
  [lindex
$auto_path_conf_list 1 ]
set camelot_cfg(pathconf_init_startdelay)
```

```
[lindex
$auto_path_conf_list 2 ]
set camelot_cfg(pathconf_resp_startdelay)
 [lindex
$auto_path_conf_list 3 ]
array set ep [ getEndpointIDs
$camelot_cfg(cam_orig_ip)
$camelot_cfg(cam_orig_port)
$camelot_cfg(ep_orig_start)
$camelot_cfg(ep_orig_end)]
puts "[getTime] :INFO: Finished localizing
Originating endpoint IDs"
puts "[getTime] :INFO: Begin localizing
Terminating endpoint IDs"
setupOrigEndpoints $camelot_cfg
(ep_orig_start)
 $camelot_cfg(ep_orig_end)
$camelot_cfg(base_dn)
$camelot_cfg(call_holdtime) $camelot_cfg
(ep_term_start)
$camelot_cfg(ep_term_end)
$camelot_cfg(orig_line_number)
$camelot_cfg(term_line_number)
$camelot_cfg
(shared_line_group_size)
puts "[getTime] :INFO: Finished configuring
Originating endpoints"
setupTermEndpoints
$camelot_cfg(ep_term_start)
$camelot_cfg(ep_term_end)
$camelot_cfg(autoanswer_delay)
$camelot_cfg(shared_line_group_siz
e)
puts "\nINFO: Finished configuring
terminating endpoints"
if { $camelot_cfg(pathconf_flag) == "1" } {
  puts "[getTime] :INFO: Configuring
  Path Confirmation"
}
```

# Chapter 14
# Appendix D

## 14.1 Statistical Details

Statistical definitions and formulae references from Matthews (2002) and Yan and Su (2009)

ANOVA stands for Analysis Of Variance. In general, this statistical procedure compares the variance of scores in each of the groups of an experiment. With the two sample t-tests, we can have experiments with only two groups. With ANOVA, we can have as many groups as we would like to have.

A one-way ANOVA has one independent variable with a number of groups. A two-way ANOVA has two independent variables. A three-way ANOVA would have three independent variables. In general, ANOVA compares the variance of scores within a group to the variance between the groups.

If the variance between groups (those treated differently) accounts for most of the difference in scores, we would say that the experimental manipulation is having an effect. The IV worked!

If most of the difference between scores is due to the variance within groups (those treated the same) we would say that the experimental manipulation is not having an effect. The IV did not work.

Note: ANOVA's can be either between subjects (different people in each group) or within subjects (the same people in each group – also called repeated measures). In this course, we will only address between subjects designs. So, we will always have different people in each group. Also note that each group can have a different number of subjects. HYPOTHESIS TESTING A one-way ANOVA is really a broad test that tests only one thing: are all the groups the same or not. If they are all the same, then the different levels of the IV had no effect (the IV did not work). Therefore, the statistical analysis is complete.

If, however, the groups are not all the same we know something is happening in the experiment. The IV had some impact; but we do not know exactly which of the groups are different, just that at least two groups are different. To determine which

S.V. Subramanian and R. Dutta, *Measuring SIP Proxy Server Performance*,
DOI 10.1007/978-3-319-00990-2_14,
© Springer International Publishing Switzerland 2013

groups are different, we do a post hoc analysis after the ANOVA test. The post hoc analysis will be discussed in the next set of lecture notes.

The nice aspect of the ANOVA hypotheses is that they are always essentially the same (note: they are always written symbolically and are never written out with words).

The null hypothesis just states that the mean of each group is the same as the others. The null hypothesis is always: H0: mu1 = mu2 = mu3 = . . . = muk

where k = the number of groups. So, there is a mu for each group.

If there are three groups the null hypothesis would be:

H0: mu1 = mu2 = mu3

The experimental (working) hypothesis is even easier. It is always, always, always:

H1: not H0

This means that the statement that all of the groups are equal is not correct. In other words, at least two groups are different.

Note: the null cannot be mu1 not = mu2 not = mu3 because this states that group 1 is different than group 2 and that group 2 is different than group 3. This is not what the ANOVA tests. It tests whether all the groups are equal or not.

For an experiment with four groups, the hypotheses would be:

H0: mu1 = mu2 = mu3 = mu4

H1: not H0

ANOVA SUMMARY TABLE The ANOVA test is always calculated using a "source of variance" summary table that looks like this: Source of Variance SS df MS F Between groups SSb dfb MSb Fobt Within groups SSw dfw MSw Total SST dfT

Note: The variance is partitioned into three rows in the summary table. One row represents the variance between groups, one represents the variance within groups and the third row represents the total – which is the sum of the two previous rows. If you know the information for two of the three rows, you can easily calculate the information for the third row using addition or subtraction. We always complete the columns moving left to right.

Sum of the squares: The SS column represents the Sum of the Squares. This is a measure of variability.

Note: The SS within is sometimes called SS error since it represents differences in scores for people treated the same. The SS total can be calculated by adding up the SS between groups and the SS within groups.

Degrees of Freedom: The df column contains the degrees of freedom. The df between groups is calculated using the formula:

df between = k − 1 where k is the number of groups.

The df within groups is calculated using the formula:

df within = N − k

where N is the total number of participants and k is the number of groups.

Note: the N value is calculated by adding up the number of subjects in each group. So, if there are three groups with ten people each, $N = 3 \times 10 = 30$.

$$Regression\,Coefficient = \hat{\beta}_1 = \frac{\sum_{a=1}^{n}(x_a - \bar{x})(y_a - \bar{y})}{\sum_{a=1}^{n}(x_a - \bar{x})^2}$$

$$Intercept = \hat{\beta}_0 = \bar{y} - \bar{x} \bullet \hat{\beta}_1 = y - x \bullet \frac{\sum_{a=1}^{n}(x_a - \bar{x})(y_a - \bar{y})}{\sum_{a=1}^{n}(x_a - \bar{x})^2}$$

**Fig. 14.1** Regression-coefficient-formula

The df total is calculated using the formula:

df total $= N - 1$

where N is the total number of subjects (Fig. 14.1).

The df total can also be calculated by adding the df between groups and the df within groups.

Note: $(k - 1) + (N - k) = N - 1$. Mean Square: The Mean Square represents the average amount of variance per degree of freedom.

The Mean Square between groups is simply:

MS between $=$ SS between / df between

The Mean Square within groups is simply:

MS within $=$ SS within / df within

F obtained: The statistic used in an ANOVA is the F statistic. The final column in the ANOVA summary table is the F obtained value. This is calculated as:

F obtained $=$ MS between / MS within

Note: The F obtained value just compares the amount of variance between groups (treated differently) to the amount of variance within groups (treated the same).

If there is large variance between groups compared to the variance within groups then the IV had an effect.

If there is large variance within groups compared to the variance between groups then the IV had no effect and the differences between groups are due to sampling error.

Note: The F obtained value is often written as:

F (df between, df within) $=$ calculated value

F critical: Once the F obtained is known, the F critical needs to be found. Remember, we are just testing to see if there is a difference between the groups, not the direction of the difference. Therefore, the F statistic is always positive. There are no one- or two-direction tests for the ANOVA. There is only one F critical value and it is always positive (Figs. 14.2 and 14.3).

$$r^2 = \left( \frac{\sum_{a=1}^{n}(x_a - \bar{x})(y_a - \bar{y})}{\sqrt{\left(\sum_{a=1}^{n}(x_a - \bar{x})^2\right)\left(\sum_{a=1}^{n}(y_a - \bar{y})^2\right)}} \right)^2 = \frac{\left(\sum_{a=1}^{n}(x_a - \bar{x})(y_a - \bar{y})\right)^2}{\left(\sum_{a=1}^{n}(x_a - \bar{x})^2\right)\left(\sum_{a=1}^{n}(y_a - \bar{y})^2\right)}$$

**Fig. 14.2** Regression-R-squared-formula

$$SE\ Residuals = \hat{\sigma}_\varepsilon = \sqrt{\frac{\sum_{a=1}^{n}(y_a - \hat{y}_a)^2}{n-2}}$$

$$SE\ Regression\ Coeff = \hat{\sigma}_{\beta_1} = \hat{\sigma}_\varepsilon \cdot \sqrt{\frac{1}{\sum_{a=1}^{n}(x_a - \bar{x})^2}} = \sqrt{\frac{\sum_{a=1}^{n}(y_a - \hat{y}_a)^2}{n-2}} \cdot \sqrt{\frac{1}{\sum_{a=1}^{n}(x_a - \bar{x})^2}}$$

$$SE\ Intercept = \hat{\sigma}_{\beta_0} = \hat{\sigma}_\varepsilon \cdot \sqrt{\frac{1}{n} + \frac{\bar{x}^2}{\sum_{a=1}^{n}(x_a - \bar{x})^2}} = \sqrt{\frac{\sum_{a=1}^{n}(y_a - \hat{y}_a)^2}{n-2}} \cdot \sqrt{\frac{1}{n} + \frac{\bar{x}^2}{\sum_{a=1}^{n}(x_a - \bar{x})^2}}$$

**Fig. 14.3** Regression-standard-error-formula

# References

Abendroth D, Killat U (2004) Numerical instability of the m/d/1 system occupancy distribution. http://www.comnets.uni-bremen.de/itg/itgfg521/

Adan I, Resing J (2001) Queuing theory. Class notes, Department of Computer Science and Mathematics, Eindhoven University of Technology, The Netherlands

Alexander AL, Wijesinha AL, Karne R (2009) An evaluation of secure real-time transport protocol (SRTP) performance for VoIP. In: Third international conference on network and system security, IEEE NSS 2009, Gold Coast

Apache Jmeter (2006) Jmeter tool. JAVA based traffic generator tool. http://jakarta.apache.org/jmeter/

Arkko J, Camarillo G, Niemi A, Haukka T (1999) Security mechanism agreement for the session initiation protocol (sip). RFC 3329, Internet Engineering Task Force. http://www.ietf.org/rfc/rfc3329.txt

Atoui M (1991) Performance measurements of the ss7/intelligent network. Military communications conference, IEEE MILCOM, McLean, vol 2, pp 774–778

Bozinovski M, Gavrinovska L, Prasad R (2002) Performance evaluation of sip based state-sharing mechanism. In: Vehicular technology conference, IEEE VTC, Birmingham, vol 4, pp 2041–2045

Cha E, Choi H, Cho S (2007) Evaluation of security protocols for the session initiation protocol. In: Proceedings of 18th international conference on computer communications and networks, IEEE ICCCN 2007, Honolulu

Chatterjee S, Tulu B, Abhichandhani T, Li H (2005) Sip based enterprise converged networks for voice/video over ip: implementation and eveluation of components. IEEE Sel Areas Commun 23(10):1921–1933

Chung MY, You JY, Sung DK, Choi BD (1999) Performance analysis of common channel signaling no. 7 based on signaling system no. 7. IEEE Trans Reliab 48(3):224–233

CISCO Internal tool (2004) Cisco camelot phone call simulation tool. CISCO proprietary tool. http://http://wwwin-vts.cisco.com/camelot/index.html

CISCO Internal tool (2006) Cisco internal automation gui environment. CISCO proprietary tool. http://http://clutch.cisco.com:9601/swims/

Cursio I, Mundan M (2002) Sip call setup delay in 3g networks. In: Proceedings of the seventh international symposium on computers and communications, IEEE ISCC, Taormina, pp 835–840

De BS, Joshi PP, Sahdev V, Callahan D (2002) End-to-end voip testing and the effort of qos on signaling. IEEE Comput 35(5):80–87

Dierks T, Allen C (1999) The tls protocol version 1.0. RFC 2246, Internet Engineering Task Force. http://www.ietf.org/rfc/rfc2246.txt

Erlang AK (1948) The theory of probabilities and telephone conversations in the life and work of A.K. Erlang. Trans Dan Acad Tech Sci 2:131–137

Eyers T, Schulzrinne H (2000) Predicting internet telephone call setup delay. In: Internet telephony workshop, IPTEL, Berlin

Fathi H, Chakraborty S, Prasad R (2004) Optimization of voip session setup delay over wireless links using sip. In: Global telecommunications conference, IEEE GLOBECOM, Dallas, vol 6, pp 4092–4096

Fink RL (2002) M/d/1 waiting line. http://bradley.bradley.edu/~rf/wait-md1.htm

Guillet T, Serhrouchni A, Badra M (2008) Mutual authentication for sip: a semantic meaning for the sip opaque values. In: Proceedings of the new technologies, mobility and security, IEEE NTMS 2008, Tangier

Gulbrandsen A, Vixie P, Esibov L (2000) A dns rr for specifying the location of services (dns srv). RFC 2782, Internet Engineering Task Force. http://www.ietf.org/rfc/rfc2782.txt

Gurabani VK (2004) Service oriented computing: enabling cross-network services between the internet and the telecommunications network. PhD thesis, Illinois Institute of Technology, Chicago

Gurbani VK, Jagadeesan L, Mendiritta VB (2005) Charecterizing the session initiation protocol (sip) network peformance and reliability. In: 2nd international service availability symposium, ISAS, Berlin

Ingolfsson A, Gallop F (2003) Queueing toolpak 4.0. Queueing ToolPak 4.0. http://www.business.ualberta.ca/aingolfsson/QTP/download.htm

Invent H (2003) Sipp tool by hp invent. SIPp Open Source SIP UA Simulator tool. http://sipp.sourceforge.net

IPTEL (2002) Sip express router by iptel. http://www.iptel.org/ser/

Johnston AB (2003) SIP: understanding session initiation protocol, vol 1, 2nd edn. Artech House Publishers, Boston

Kaji T, Hoshino K, Fujishiro T, Takata O, Yato A, Takeuchi K, Tesuka S (2006) Tls handshake based on sip. In: International conference on computer science and information technology, IEEE ICSIT 2006, Wista

Kim J, Yoon S, Jeong H, Won Y (2008) Implementation and evaluation of sip-based secured voip communication systems. In: International conference on embedded and ubiquitous computing, IEEE EUC 2008, Shanghai

Koba EV (2000) An M/D/1 queuing system with partial synchronization of its incoming flow and demands repeating at constant intervals, vol 36. Springer, Boston

Kohn PJ, Pack CD, Skoog RA (1994) Common channel signaling networks: past, present, and future. IEEE J Sel Area Commun 12(3):383–394

Lazar AH, Tseng K, Lim KS (1992) Delay analysis of singapore national ccs7 network under fault and unbalanced loading conditions. In: Proceedings of the international conference of complex systems 1992, IEEE ICCS, Singapore

Lennox JM (2004) Services for ip telephony. PhD thesis, Columbia University, New York

Lennox JM, Rosenberg J, Schulzrinne H (2001) Common gateway interface for sip. RFC 3261, Internet Engineering Task Force. http://www.ietf.org/rfc/rfc3050.txt

Lennox J, Schulzrinne H, Wu X (2004) Cpl: a language for user control of internet telephony services. RFC 3880, Internet Engineering Task Force. http://www.ietf.org/rfc/rfc3880.txt

Lin YB (1996) Signaling system number 7 unknotting of how all linking and switching works. IEEE Potentials, 15(3)

Malas D (2009) Sip end-to-end performance metrics.txt. Internet-draft, Internet Engineering Task Force. http://tools.ietf.org/html/draft-malas-performance-metrics-03

Matthews PG (2002) Statistical tool excel. Linear regression analysis. http://www.experts-exchange.com/articles/Microsoft/Development/MS_Access/Simple-Linear-Regression-in-MS-Access.html

Mealling M, Daniel R (2000) The naming authority pointer (naptr) dns resource record. RFC 2915, Internet Engineering Task Force. http://www.ietf.org/rfc/rfc2915.txt

Rajagopal N, Devetsikiotis M (2006) Modeling and optimization for the design of IMS networks. In: Proceedings of 39th annual simulation symposium 2006, IEEE ANSS06, Huntsville

Ram KK, Fedeli IC, Cox AL, Rixner S (2008) Explaining the impact of network transport protocols on sip proxy performance. In: Proceedings of international symposium on performance analysis of systems and software, IEEE ISPASS 2008, Austin

Rosenberg J, Schulzrinne J, Camarillo H, Sparks PJ, Handley R, Schooler E (2002) Sip: session initiation protocol. RFC 3261, Internet Engineering Task Force. http://www.ietf.org/rfc/rfc3261.txt

Schulzrinne H, Narayanan S, Lennox J, Doyle M (2002) Sipstone – benchmarking sip server. http://www.sipstone.org/files/sipstone_0402.htm

Stewart WJ (2003) Probability, queuing models and markov chains: the tools of system performance evaluation. Class notes, Department of Computer Science, North Carolina State University, Raliegh

Subramanian SV, Dutta R (2008) Comparative study of $m/m/1$ and m/d/1 models of a sip proxy server. In: Australasian telecommunications networking and application conference, IEEE ATNAC 2008, Adelaide

Subramanian SV, Dutta R (2009a) Measurements and analysis of m/m/1 and m/m/c queuing models of the sip proxy server. In: Proceedings of the 18th international conference on computer communications and networks, IEEE ICCCN 2009, San Francisco

Subramanian SV, Dutta R (2009b) Performance and scalability of $m/m/c$ based queuing model of the sip proxy server – a practical approach. In: Australasian telecommunications networking and application conference, IEEE ATNAC 2009, Canberra

Subramanian SV, Dutta R (2010a) Comparative study of secure vs non-secure transport protocols on the sip proxy server performance: an experimental approach. In: Second international conference on advances in recent technologies in communication and computing, Submitted to IEEE ARTCom 2010, Kottayam

Subramanian SV, Dutta R (2010b) Performance measurements and analysis of sip proxy servers in local and wide area networks. In: Second international conference on advances in recent technologies in communication and Computing, Submitted to IEEE ARTCom 2010, Kottayam

Sulaiman N, Carrasco R, Chester G (2008) Impact of security on voice quality in 3g networks. In: Industrial electronics and applications, IEEE ICIEA 2008, Singapore

Sun Microsystems (2003) Sip servlet api. Java Community Process, JSR 116. http://java.sun.com/products/servlet/

Swart H, Verschuren H (2002) Java applet for Queuing Theory-Queuing Simulation tool, version 2.0 http://http://www.win.tue.nl/cow/Q2/

Towsley D, Alouf S, Nain P (2001) Inferring network characteristics via moment-based estimators. In: Twentieth annual joint conference of the IEEE computer and communications societies, IEEE INFOCOM, Anchorage

WANem tool (2008) Wan emulator tool. WAN Emulator tool software. http://www.wanem.sourceforge.net/

Wang JL (1991) Traffic routing and performance analysis of common channel signaling system no. 7 network. In: Global telecommunications conference, IEEE GLOBECOM, Phoenix, Arizona

Wireshark (2008) Wireshark protocol analyzer tool. Wireshark Protocol Analyzer tool. http://www.wireshark.org/

Wu WC (1994) Performance of common channel signaling no. 7 protcol in telecommunications network. In: Proceedings of the 37th midwest symposium on circuits and systems, Lafayette, vol 2. IEEE, pp 1428–1431

XR Perfomon tools (2006) Perfmon tool. Perfmon tool software. http://www.xratel.com/perfmon.asp

Yan X, Su X (2009) Linear regression analysis: theory and computing, vol 1, 2nd edn. World Scientific, Singapore

Yanik T, Kilinc H, Sarioz M, Erdem S (2008) Evaluating sip proxy servers based on real performance data. In: Symposium of performance evaluation of computer and telecommunication Systems, IEEE SPECTS 2008, Edinburgh

Printed in the United States
By Bookmasters